A NOTE ON THE AUTHORS

Richard Evershed FRS is Professor of Biogeochemistry at the University of Bristol. He researches the diets and agriculture of our ancestors and the environmental impacts of modern farming. His pioneering analytical chemical methodologies are widely used. In the area of food fraud his team developed a method for detecting the adulteration of vegetable oil. He believes in wholesome food and the supplier's responsibility to ensure its quality and authenticity, especially given *'we are what we hope we're eating!'*

Nicola Temple is a biologist, conservationist and science writer. Based in Bristol, Nicola works with universities, research councils and individuals to develop engaging science stories on the impact of research beyond the closeted world of academia. It was while investigating the impact of Richard's work that Nicola was first introduced to the world of food fraud.

Also available in the Bloomsbury Sigma series:

Sex on Earth by Jules Howard
p53: The Gene that Cracked the Cancer Code by Sue Armstrong
Atoms Under the Floorboards by Chris Woodford
Spirals in Time by Helen Scales
Chilled by Tom Jackson
A is for Arsenic by Kathryn Harkup
Breaking the Chains of Gravity by Amy Shira Teitel
Suspicious Minds by Rob Brotherton
Herding Hemingway's Cats by Kat Arney
Electronic Dreams by Tom Lean
Death on Earth by Jules Howard
The Tyrannosaur Chronicles by David Hone
Soccermatics by David Sumpter
Big Data by Timandra Harkness
Goldilocks and the Water Bears by Louisa Preston
Science and the City by Laurie Winkless
Bring Back the King by Helen Pilcher
Furry Logic by Matin Durrani and Liz Kalaugher
Built on Bones by Brenna Hassett
My European Family by Karin Bojs

SORTING THE BEEF FROM THE BULL

THE SCIENCE OF FOOD FRAUD FORENSICS

Richard Evershed
&
Nicola Temple

BLOOMSBURY
sigma

Bloomsbury Sigma
An imprint of Bloomsbury Publishing Plc

50 Bedford Square
London
WC1B 3DP
UK

1385 Broadway
New York
NY 10018
USA

www.bloomsbury.com

BLOOMSBURY and the Diana logo are trademarks of
Bloomsbury Publishing Plc

First published 2016. Paperback edition 2017

British Library Cataloguing-in-Publication Data
A catalogue record for this book is available from the British Library.

Library of Congress Cataloguing-in-Publication data has been applied for.

ISBN (paperback) 978-1-4729-1135-3
ISBN (ebook) 978-1-4729-1134-6

2 4 6 8 10 9 7 5 3 1

Chapter illustrations by Nicola Temple
Chemical structures by Richard Evershed

Typeset by Deanta Global Publishing Services, Chennai, India
Printed and bound in Great Britain by CPI Group (UK) Ltd, Croydon CR0 4YY

To find out more about our authors and books visit www.bloomsbury.com.
Here you will find extracts, author interviews, details of forthcoming
events and the option to sign up for our newsletters.

Contents

Foreword by Chris Elliott   6

Introduction: Not Eggsactly What You Thought   9

Chapter 1: Food Fraud 101   17

Chapter 2: Busting the Food Cheats   49

Chapter 3: A Slippery Deal   85

Chapter 4: Hake Today, Cod Tomorrow   107

Chapter 5: What's Your Beef?   139

Chapter 6: Milking It   173

Chapter 7: Seasoned Criminals   201

Chapter 8: Bogus Beverages   225

Chapter 9: White Collard Crimes   255

Chapter 10: Thoughts for Digestion   273

Appendix: Some of the Chemical Structures   289

Notes   295

Acknowledgements   307

Index   311

Foreword

To many, food fraud is about trivial acts of cheating undertaken by a few dodgy butchers or fast-food outlets. Perhaps it is a handful of sawdust popped into the sausage mix, or a little something added to the lamb to eke it out a bit. While there is no doubt such acts occur fairly frequently and they certainly shouldn't, food fraud is really much more complex, sinister and organised, and it has the potential to ruin businesses and the lives of those affected. I have had personal contact with a number of food-business operators who have told me their biggest dilemma is to decide if they should cheat in the same way as their competitors, or go out of business. I am also acutely aware that companies who 'do the right thing' can lose contracts to those who clearly don't.

For as long as food has been prepared and sold there has been cheating. Looking at the history of food fraud is a wonderful way to explore the darker side of human nature. There have been scams that have poisoned people and scams that have killed people, but most often the fraud goes undetected. This is the 'business model' of the fraudster, of course. The longer he or she, or in many cases 'they', get away with it, the more profit they make from their deception. My own view of the horse-meat fraud, based on intelligence I was able to source, was that it had been going on for at least five years before a random test alerted the nation to the scale of the problem. I uncovered a scam with oregano in the UK recently, and a few reliable sources have told me that this had been going on for a lot longer.

Food fraud becomes food crime when the level of organisation increases to the point that formal or informal networks of perpetrators are involved, and have differing roles in the criminal activity. Firstly, there is the 'fraud inventor', i.e. the person (or persons) who devises the way of cheating; there are some excellent examples of the ingenuity of such deceptions in this insightful book. Secondly, there are those that deal with the logistics – the people who organise the transportation of the frequently produced goods across countries, borders and, in some cases, across the globe. Thirdly, there are those that develop the countermeasures to detection working out how to evade laboratory testing and auditing. The final 'expert' of the gang is often the enforcer, those that threaten and bully the most vulnerable in the food industry to turn a blind eye or become complicit in the fraud itself. I have met some of these at first hand, and know the degree of menace that comes with them. Would I be brave enough to stand up to them if I was a small food company that realised the police wouldn't be overly bothered about a food fraud issue? I have often wrestled with this thought, and on a good day believe I might.

Sorting the Beef from the Bull gives a fascinating insight into many of the complexities I have referred to. The examples selected should enable the reader to have a much better understanding of the driving forces behind food fraud, the lengths the cheats will go to in order to make a profit and the ways in which science is trying to help deliver the means of introducing more deterrents to one of the world's fastest-growing industries. Barely a day goes by now without news of another scam cropping up somewhere in the world. Worryingly, quite a lot of them could impact the UK in a way that would make Horsegate feel like hors d'oeuvres.

Professor Chris Elliott
Pro Vice Chancellor, University of Belfast

Not Eggsactly What You Thought

Bird eggs are perfect structures that contain all of the materials needed to support the growth of a new life or, in many unlucky cases, provide a highly nutritious meal to any number of different critters. It takes 25 hours for an egg to form within a female bird; one hour for the microscopic egg cell to attach to a globule of yolk and pass through the oviduct, where it may (or may not) become fertilised, and then reach the lime-producing section of its journey. There it will sit for 24 hours as it is coated with a hard shell to protect its precious contents. The finishing touch is a layer of pigment to help camouflage the egg, which is added just before it is laid – much like a wax spray that is applied as a car leaves the car wash. It's truly a remarkable process.

Unfertilised versions of these little packages, mostly of the chicken variety, have become an important part of our diet. They are the basis of the full English breakfast, the foundation for a slathering of hollandaise sauce (not to mention part of the hollandaise sauce itself), and when the whites are whipped into a frenzy with sugar they become the most delicate of meringues.

Now, what if we told you that we could emulate that perfect little package in a fraction of the time the chicken takes to make it? All you need is a few ingredients that can be purchased online. First, take some sodium alginate. This comes from brown algae and when it binds with water it forms a viscous gum. It's commonly used in the food industry to increase viscosity and also largely responsible for those gelatinous chunks that you find in wet cat food. Mix this sodium alginate with water and stir for about one and a half hours. Then add some gelatin, which you can buy at your local supermarket for making trifles, jams or yogurt, and mix this in thoroughly. Leave it for about 10 hours until any bubbles have disappeared. Then add in some sodium benzoate and alum – both widely used food preservatives. Maybe throw in some lactones if you have them (if not, they're available online), as these are the compounds that contribute to the aroma of butter, cheese, eggs and other foods. This is our base substance and it can be made well in advance.

Next, put some of this mixture into a bowl and add a little colouring agent – something labelled 'orange red powder' might work. This is your yolk mixture and you can now pour it into yolk-shaped moulds. Dip these moulds briefly into edible calcium chloride. This is a permitted food additive that's used to firm up soybean curds for tofu and is commonly used as an electrolyte in sports drinks. This will just help to coagulate those yolks so they stay together nicely.

Pour a portion of the remaining 'egg white' mixture into a plastic mould, add your 'egg yolk' and dip that whole thing into the calcium chloride again to stabilise it all. Finally, pluck out your firm gelatinous egg and dip it into a mixture of melted paraffin wax and gypsum powder (plaster) to create that wonderful hard shell. Simple, really, and in a 25-hour period you can pop out way more eggs than a chicken can! They can be placed into egg cartons and sold at the market and profits are double to quadruple those of traditional chicken farmers. Oh, and a little bonus is that any leftover 'egg white' mixture can be dyed green or purple and with a little added juice (just a splash), you've got yourself some fake grapes as well! No chicken can do that.

Fake eggs started appearing in China in the mid-1990s and continue to crop up from time to time. The fakes are so good that people are cooking them up and eating them. There's loads of information online to show people how to tell the fakes from the real thing (hint: the fakes don't have a shell membrane – that thin layer of skin below the shell). The health consequences of eating these eggs are unclear; though most of these ingredients are already used in food products, they are not used in the quantities found in the fake eggs.

Given the title of the book, you probably expected us to jump immediately into the 2013 European horse meat scandal. Perhaps start with some groaner joke or clever pun about eating horse meat. Sorry to disappoint, but we had to *rein* in that type of humour as we thought it could all be a *bit* much really. Besides, in terms of creative food fraud we thought slipping some horse into a batch of minced beef wasn't nearly as imaginative as creating an entire egg from a bunch of powders and liquids. Although we have to admit that the horse meat scandal did *spur* us into writing this book (groan ... OK, we're done ... promise).

Building eggs from chemical compounds is, as far as we're concerned, an extreme example of food fraud. It's a carefully manufactured replica of the real thing with only one purpose – to fool trusting people into buying it in order to earn the perpetrator some money. Yet we process food to replicate other things all the time. There are soy-based products that taste and act like cheese. Fish paste is flavoured and fashioned to look like crab. A week-old cut-up apple has been dipped in additives to make it look like a freshly cut apple. And gelatin, sugar and some other additives are mixed together into fried-egg-shaped sweets for the delight of children. So why is one fake egg acceptable while the other is not? Because it is less about what we do to the food and more about what we say on the packaging.

Food science has allowed us to store, ship and conveniently whip up meals like never before. It has led to a diverse alimentary landscape and revolutionised food in ways we never thought possible – ways most consumers may still not realise are possible, in fact.

As an analogy, let's consider for a moment how computer-generated imagery (CGI) has transformed cinema. It has enabled the lines between reality and fiction to be blurred so effectively that viewers aren't always sure where the stunt person ends and the CGI effect begins. In the right hands this is movie magic, but in the wrong context it can be used to deceive viewers. The same is true of our food. Modern food undergoes a lot of behind-the-scenes processing, which can warp a consumer's understanding of reality. Our existing food supply has created a perfectly legal and legitimate system that has many blurred lines – grey areas if you will. And it is within these shadows that criminals can commit food fraud.

When we began the research for this book on food fraud we very soon became overwhelmed. Food fraud is, quite frankly, an enormous topic – it could fill many books, some of which would be far more interesting than others. There is an entire story about the political and legal structure that seems to both prevent and enable food fraud. Another story lies in the economics of it all – the cost to industry, the cost to consumers and the financial benefits for the fraudsters themselves. Then there are the non-economic costs of food fraud – the environmental and social costs as well as the threat to human health. Another important story is how food fraud is linked to other criminal activities including drug trafficking, tax evasion, illegal immigration and even slavery.

It wasn't long into this journey that we started to truly question how and where we buy food, the integrity of our food supply, the benefits and drawbacks of a global supply network and what our most basic values are around food. There are many books and as many movements that are looking at alternative systems to help build resilience, sustainability and integrity into our food supply. While this is a fascinating discussion, we didn't feel this was our story.

We are science geeks (well, one of us is an internationally renowned analytical organic chemist and Fellow of the

Royal Society and the other is a geek). We felt that the science story was by far the most interesting and perhaps also the most overlooked in the popular media. So while we necessarily touch on economics, health, legislation and politics in each chapter, we have largely lumped much of this discussion into Chapter 1 so we would have more room to talk about the science in the rest of the book.

The science of detecting (and committing) food fraud has evolved over the centuries. In Chapter 2 we discuss some of the first scientists to tackle food fraud and the basic tools they had to work with. Their discoveries provided the first evidence of the prevalence of food fraud and this ultimately helped to shape the first food policies on both sides of the Atlantic.

However, since those first policies were put in place, our food networks have become complex and globalised. These days, our food whizzes around the planet, getting processed and mixed into ready-meals that no longer bear any resemblance to living things that once grew. We have created an ideal environment for criminals to operate within our global food system. And now we must face the challenge this creates in terms of relying on labels to tell us what this unrecognisable food is and developing tests to confirm that what the label says is true – authentication.

In Chapter 2 we also look at how the scientific method is applied to first frame and then test questions of food authenticity. We put all of this into practice by working through a case study with one of the purest food substances on the planet: honey. Let's face it, it's bee barf in a jar with very little human processing required – it's the perfect simple food. Yet, honey often finds itself in sticky scandals. As we worked through the example of authenticating the different label claims on a hypothetical jar of honey, it became clear to us (and hopefully it will to you too) that this is truly a game of cat and mouse between the scientists detecting fraud and the criminals committing it. If scientists develop a new test to detect corn syrup in honey, the fraudsters will find a new type of syrup that escapes this new detection method. It is a continual evolution and will always be so.

As we work through the honey example, we introduce many of the analytical techniques that come up again in later chapters – in this way, it is a bit of a 'go-to' chapter in terms of the methods. However, if we couldn't use the method in our honey example, we saved it for where we could work it into another example. And this is exactly what we do over the subsequent chapters. We dedicate Chapters 3–9 to foods that are most vulnerable to food fraud. And as we wrote these, we were equally awed by the sophistication (and audacity) of some of the scandals committed and by the cleverness of the methods used to catch them.

We discuss vegetable oil in Chapter 3 – a foodstuff that has a long history of adulteration and yet continues to present modern-day challenges in terms of authentication. Vegetable oil is one of the most adulterated products on the market, with cheap oil frequently added to expensive oil to make it go a little further. Such substitutions are impossible for us to discern as consumers, and surprisingly difficult to distinguish for the food analyst. Vegetable oil has been at the root of swindles that have shaken the US economy and caused hundreds of deaths in Spain; it's a slippery (and sometimes deadly) deal.

From oil we move to the gross mislabelling of seafood in Chapter 4. We look at how DNA-based methods are trying to overcome the challenge of identifying the most speciose group of vertebrates on the planet. There are over 30,000 described species of fish and many of the commercially important species have more aliases than Simon Templar – Atlantic cod has 56 English names used in Canada alone! This creates confusion as these species move between countries – first to be processed into characterless fillets and then to be sold to consumers. While the optimistic viewpoint is that this is simply a matter of mistaken identity, the fact that it is nearly always cheaper species being substituted for more expensive ones suggests a more malicious intent.

Of course, we can't ignore horse meat all together (it's on the cover after all). Chapter 5 explores the methods used to identify what's really in our burgers as well as our curries, kebabs and chicken breasts. We share cases of putrid meat being redirected back into the human food chain, 'all beef' sausages that would be more accurately labelled 'all *but* beef' and ham and chicken that's just a lot of added water. If you're vegetarian and think you can skip this chapter, think again. There are cases of meat adulterating our spices and blood products making their way into our baked goods. This chapter isn't just for carnivores.

From the melamine milk scandal in China to fake milk made of urea and shampoo in India, in Chapter 6 we discuss corruption in our dairy products, including cheese and butter. We discuss how genuine quests to replace many animal-based fats with vegetable-based products has helped equip fraudsters with the tools they need to adulterate our dairy foods. Be warned that some of the examples may *churn* your stomach.

Shortly after the horse meat scandal in 2013, another food scandal hit the headlines in both North America and Europe. Almond and nut protein was discovered in cumin and paprika. This scandal didn't seem to make as many headlines as horse meat in the UK, yet it had far greater potential for causing human harm in terms of allergens. Above all, this spicy scandal illustrated how a single substitution – one fraudulent act – can permeate through the food supply. In Chapter 7 we look at the seasoned criminals who use ground spices to hide all manner of cheap adulterants and how analysts go about finding them.

The motivation behind all of the scandals discussed in this book is money. And there is no greater example of the economic gains that can be made by falsifying a food product than in the wine industry. The fine and rare wine market is rife with scandals, but the criminals operating in these circles are often as wealthy as their victims and they have the refined palates and product knowledge that enable them to carry off such scams. In this world, as we explain in Chapter 8, it is more about product knowledge and less about chemical analyses.

In Chapter 9 we look at all things wholesome and good – fresh fruit and vegetables as well as some grains, cereals and pulses (lentils) for good measure. Yet even these seemingly healthful products can't escape food fraud. While there aren't any examples (yet) of fabricated tomatoes or parsnips posing as carrots, our fresh fruit and veg is being misrepresented in the way it is preserved, produced and processed. From mangoes sprayed with formalin (a mixture of water and formaldehyde used to preserve tissues) to innovative new methods that keep food looking fresh well beyond what we think is possible, we dip into the grey areas of food production that are legal on paper, but feel deceitful in practice.

Finally, we look at what we can do to reduce our vulnerability to food fraud. As consumers trying to nourish ourselves in an incredibly diverse and complex alimentary landscape, we are forced to trust the information that's on product labels. Whether we are meant to actually understand what's on the label is an entirely different story, of course. Yet we don't want to leave you feeling hopeless (as so many of our editors who never got to read the last chapter did) and there are indeed actions we can take. You've taken the first step, which is to open the cover of this book and read this far. We hope that by taking the second step (reading the rest) you will end up a more informed consumer by knowing what is possible in the world of food fraud – both in terms of how it is committed, how it is detected and how to avoid it.

CHAPTER ONE

Food Fraud 101

Adulterate – a word similar to adultery and arguably as scandalous.

In 2008, approximately 52,000 children in China were hospitalised owing to the adulteration of milk powder with the nitrogen-rich organic compound, melamine. Six children died. Yet a Google search on this scandal brings up only a quarter of the results that 'Ashton Kutcher adultery' does. It would seem as though our society is far more interested in the alleged affair(s) of a celebrity than an event affecting thousands of people – innocent babies – involving one of the United States' biggest food trade partners. Is this a reflection of our insatiable appetite for celebrity gossip or our apathy towards the food we eat? It's probably both ... and perhaps it's time we refined our tastes.

Defining the problem

'Food adulteration' means to reduce the quality of a product by substituting or adding something cheaper. Adulteration is most often motivated by the desire to make a quick buck and in 2009 the US Food and Drug Administration (FDA) defined economically motivated adulteration as 'the fraudulent, intentional substitution or addition of a substance in a product for the purpose of increasing the apparent value of the product or reducing the cost of its production, i.e., for economic gain'.[1] Bee Wilson, in her book *Swindled*, broke this down into two very simple principles: poisoning and cheating.

The United Kingdom's Food Standards Agency (FSA) estimates that about 10 per cent of the food we buy from supermarket shelves is adulterated, but this is really just an educated guess. In 2010, the Food Safety and Standards Authority of India (FSSAI) conducted a national study that involved the analysis of 117,000 food samples collected from all over India. They found that about 13 per cent of samples had been adulterated, but that the adulteration rates could be as high as 40 per cent in some locations (Chhattisgarh) and as low as 4 per cent in others (Delhi). Few other countries have done such a wide-scale assessment of adulteration.

Adulteration is just one of a number of ways we can be cheated in what we eat, though it is perhaps the most common. After all, some beef or chicken in a lamb kebab or some cheaper variety of rice labelled as basmati can bump up profits with relatively low risk of being detected. But food fraud is diverse in its methods of poisoning and cheating. Food that is past its 'use by' or 'best before' date can be deceitfully repackaged and sold. Animal by-products can be recycled back into the food chain. Products can be purposely mislabelled in terms of their species or country of origin in order to fetch a higher price. False statements can be made with regard to the production of the product, such as claims that eggs are free-range, produce is organic or salmon are wild-caught. These are all acts of fraud that earn someone somewhere along the food chain easy money, while others have to pay the price.

Misrepresentation of food in any way is morally wrong. As consumers, we are at the mercy of labels and if they are misleading or untrue, we are at risk. Some producers go to great lengths to create a genuine premium product, potentially with superior taste and/or nutritional benefits. Conscientious consumers are likely to be willing to pay more for such products. But when a product is misrepresented, consumers are being cheated out of money, they are being cheated out of nutrition and the legitimate producers are being cheated out of business.

Estimates of what food fraud costs the global food industry each year range from £6.4 billion to £31 billion (US$10 billion to $49 billion), depending on the source. Food fraud can drive down prices across the board as businesses try to compete with the price on a fraudulent product. When fraud is detected it can force the industry to conduct rigorous testing on a wide variety of products and potentially recall products; companies can begin to lose sales and can ultimately go bankrupt. The melamine milk incident in 2008 forced China's dairy giant, Sanlu Group Co. Ltd, into bankruptcy by December of that year. More than 30 milk brands were affected globally by the adulterated milk, forcing more than 60 countries to ban or recall Chinese dairy products at an estimated cost of £11.5 billion (US$18 billion).

In 2007, leading US pomegranate juice manufacturer, POM Wonderful, sued the small California-based beverage company, Purely Juice Inc., for falsely advertising one of its products as 100 per cent pomegranate juice. Suspicious of Purely Juice's low prices, POM Wonderful secured tests from seven independent laboratories that confirmed that the Purely Juice samples contained little or no pomegranate juice and consisted mainly of corn syrup and other fruit juices. Purely Juice claimed a supply chain issue, stating that they were unaware that their supplier of 100 per cent concentrate was not providing the real thing. In 2008, Purely Juice, with annual revenues of approximately $10 million (£6.5 million), was ordered to pay nearly $2 million (£1.3 million) to POM Wonderful for damages, disgorgement of wrongful profits and legal costs. Purely Juice appealed against the decision, but in 2010 the higher court affirmed the lower court's decision and the president and founder of Purely Juice, Paul Hachigian, was held personally liable. Purely Juice is now out of business. The economic costs can be acute.

The costs of food fraud go well beyond a company's bottom line. There are also social, environmental and health costs associated with these practices. There have been many cases throughout the UK now where kebabs and curries have been labelled as lamb or chicken, but have largely contained cow or pig. You will even learn in

Chapter 5 that raw chicken breasts have been adulterated with cow and pig protein. This presents an enormous challenge to millions of people whose religious beliefs don't allow the consumption of cows or pigs, including the UK's second largest religious group, Muslims.

Some of the environmental consequences of food fraud have been brought to light by the US organisation Oceana. Between 2010 and 2012, Oceana collected more than 1,200 seafood samples from 674 retail outlets in 21 states to determine whether products were being labelled accurately. They found that 33 per cent of these samples were mislabelled. Among other consequences, this dishonest labelling of species can undermine sustainable fishing practices and conservation efforts. First, it creates a market for illegal fishing, including the capture of at-risk species. Second, it misleads consumers into thinking that certain species are readily available, which may not necessarily be the case. Red snapper is a commercially important species that has had harvest restrictions put in place to help conserve it. Yet, if it's on every menu, the perception is that the species is abundant, when in fact what's really on the menu is farmed tilapia. Third, consumers who are trying to make sustainable seafood choices using the various consumer guides, such as the Marine Conservation Society's *Good Fish Guide* or the Monterey Bay Aquarium's *Seafood Watch*, may not be buying the sustainable product they think they are. Instead, they may be contributing to the overfishing of vulnerable stocks (more in Chapter 4).

It is human health, however, that arguably carries the greatest cost when it comes to food fraud. An estimated 300,000 children fell ill in the 2008 melamine scandal in China. As we mentioned earlier in this chapter, over 50,000 of these children were hospitalised and six died (we'll revisit this in Chapter 6). In 1981, more than 1,000 deaths and 25,000 serious injuries in Spain were attributed to olive oil that had been adulterated with industrial quality rapeseed (canola) oil. Subsequent investigations suggest that tomatoes heavily dosed with pesticides may have been the true cause of this Spanish outbreak, though oil remains the official culprit (more on this in Chapter 3). In 1986, eight Italians died and 30 were hospitalised when they drank wine adulterated with methyl alcohol, an ingredient of antifreeze and solvents. Austrian wine has been adulterated with diethylene glycol (DEG), the major component of some brands of antifreeze (we discuss this in Chapter 8). There are reports of phoney tofu cakes made from gypsum, paint and starch being sold in Shanghai, mystery meats in London curries and milk adulterated with urea in India. When you begin to delve into the many ways our food is adulterated it is surprising there aren't more deaths and illnesses associated with food fraud.

Of course, the stories captured by the media are only the acute cases. When a large number of people fall ill in a short period of time it's easier to track the source. What we can't even begin to comprehend, however, are the long-term health consequences of substances that slowly accumulate within our body tissues. The long-term health effects of using plasticisers as clouding agents in fruit juices (Chapter 8) and jams, adding the illegal colour additive metanil yellow to turmeric (Chapter 7) or the antibiotic residues in unregulated meats are, as yet, unknown.

It's necessary to put the human health concern into perspective though. Approximately 48 million Americans become ill from foodborne diseases, such as *Salmonella*, each year. About 128,000 of these people will end up in hospital and 3,000 will die. In the UK, these numbers are considerably lower: about one million people suffer from foodborne illnesses each year, 20,000 of them will receive treatment in hospital and there will be around 500 deaths. Similar estimates for food fraud-related illnesses simply don't exist, but it is no doubt several orders of magnitude less. The difference, however, is that foodborne illnesses are generally a result of accidental contamination and poor food-handling regimes. Fraud, on the other hand, is intentional. Someone, somewhere in the food supply chain has made a conscious decision that puts others at risk, and this is very alarming if you consider that this could be done with the intent to harm rather than simply to make a quick buck.

The fraud forecast: cloudy with a chance of harm

This was the motivation behind the US launching the National Center for Food Protection and Defense (NCFPD) – a Homeland Security Center of Excellence – in 2004. After 9/11 the US took a closer look at its points of vulnerability. Food came out as an obvious threat to homeland security as people consume it multiple times per day and the supply chains can be long, international and somewhat anonymous due to their complexity. The NCFPD was set up as a multi-disciplinary group of researchers to improve detection of food adulteration and to address the vulnerabilities in the US food system in terms of where it could be prone to attack with the intent to harm.

Researchers at the NCFPD are looking for ways to bring together data so that it can help to make predictions. The idea is that data would be gathered from a number of different sources and analysed to look for factors that might spark incidents of adulteration. For example, a report of prolonged drought in Spain that has led to low olive yields, combined with a history of adulteration in the olive oil business, would raise a red flag.

'We speak a lot about triggers,' said Dr Amy Kircher, Director of the NCFPD, 'and this can be climate change or even newly touted health benefits of a particular product. For example, when the health benefits of pomegranate were being heavily advertised, we saw a dramatic increase in the number of pomegranate products out there. But the US was still producing the same number of pomegranates. Production hadn't increased. How is the gap being filled?'

The idea is that the food fraud forecast would be used to inform decision-making, rather than leading to complete closures for industries. 'It is still looking for a needle in a haystack in terms of finding adulterated products,' said

Kircher, 'but this tool would at least allow us to say, "don't look at the whole haystack, just look in this bottom corner".'

Even knowing which corner of the haystack to look in might give regulators the edge they need to move towards a proactive rather than a reactive strategy. Prior to the European horse meat scandal – referred to now in the UK as Horsegate – the US didn't conduct DNA testing on their minced beef. Now they do. The adulterators will now be looking to identify the next weakness in the system.

In fact, Kircher refers to the adulterators as 'intelligent adversaries'. After all, food fraud can be big business and while scientists on one side are finding sophisticated techniques for detecting the adulterated food, there are scientists on the other side developing equally sophisticated techniques for adulterating it. It is a cat and mouse game – an arms race between scientists and fraudsters – and by its very nature it will always be so.

Traditions of trickery

History certainly would suggest this will always be the case; food fraud is not a new phenomenon. Ten thousand years ago, as our farming ancestors began producing surpluses that could be traded, the possibility for food fraud emerged. People learned to grind grains, ferment fruit to preserve it and turn milk to butter and yogurt to make it more easily digestible; these were the most basic of processed foods. Ingenious prehistoric entrepreneurs, recognising the enhanced value of these premium and specialist goods, would have realised there was a profit advantage in replacing more expensive foodstuffs with cheaper ones. The simple act of adding water to milk could easily have gone undetected by the recipient, since the change in appearance and taste would have been unobservable. But the profit margin for the producer was improved. Indeed, detecting added water in food remains one of the major challenges of the food fraud field, and it's a theme we will return to in later chapters.

And the fraud continued. In ancient Rome, lead was added to soured wine to mask the flavour. It has been argued that this was the reason many wealthy Romans were sterile as well as mentally incompetent.[2] In the eighteenth century, alum, a colourless compound used in dyeing and tanning, was used in bread to make it appear more fashionably white. It was common practice for storekeepers to have two sets of weights behind the counter – one for buying and one for selling. Copper sulphate gave pickles a more appealing though utterly unnatural green vibrancy, which meant they fetched a better price than their pallid pickled shelf-mates. Leaves were picked from sloe shrubs, which commonly grow along British hedgerows, and boiled with the poisonous compound copper acetate to be passed off as green tea. Beer was adulterated with *cocculus indicus*, a drug that caused convulsions due to the active ingredient picrotoxin. This provided the drinker with a somewhat more inebriating effect in order to hide the fact that the brewers had been sparing with the hops and malt. In fact, food fraud was so widespread during the nineteenth century that one *New York Times* journalist wrote in 1872 that 'most people come to accept it nowadays as practically inevitable' and added that it was best people didn't really know 'what abominable messes they are constantly putting down their throats under the most innocent disguises'.[3] Indeed, best not to know what scrapings from the floor have been added to your premium black pepper.

Today's food fraud may be more sophisticated, but it is certainly no less pervasive. Just in the time we've been writing this book, there have been at least five scandals in UK headlines: nut protein in spices, rat sold as mutton in China, unidentified meat in curries, goat's cheese made from sheep milk and cheap oriental perch sold as sea bass. The horse meat scandal heightened awareness of the issue and one doesn't have to look too far before yet another scandal is uncovered.

Tipping the scales on fraud

Behind every decision to commit a fraudulent act there are costs, benefits and motivators. On one side of the scale there is a reward (usually money) and a trigger (being undercut by competitors). On the other side of the scale there is effort, risk and one's ability to sleep at night (guilt). Each person's sense of guilt is probably well established, but if the scales can be tipped so that effort and risk far outweigh the rewards and triggers, we stand a chance of reducing food fraud.

Money is a powerful reward and unless it miraculously decreases in importance, it will continue to be so in the future. With the exception of fraudulent acts that are intended to harm, the goal behind all other food fraud is economic gain. This can be as simple as someone seizing a one-off opportunity to make a little extra cash by watering down some freshly squeezed juice at a market stall, right up to organised criminal networks turning over huge annual profits (fraud can be big business). In fact in 2015, Europol reported that criminal networks are increasingly turning to the food and beverage industry as an arena to conduct their 'business'. But at the root of it all – whether it's small crime or big crime – lies greed, and there are strong biological and social factors that drive us to want more resources than our neighbours. An olive grower in Italy, struggling to keep the family business alive, who watches nearby growers get rich by adulterating their oil with cheaper oils must be tempted to make their own product stretch a little further too. It takes a strong moral compass to resist temptation under such conditions. Desperation can steer people in unexpected directions.

In the mid-nineteenth century, 2.5 million hectares (6.2 million acres) of French vines were destroyed by *Phylloxera*, a relative of the aphid with a ravenous appetite for the sap of grape vines. Desperate times called for desperate measures and vast amounts of raisins were imported from Greece to create raisin wines. Some *vigneron*s prepared bottles of wine with little or no grape products at all, going for a laboratory approach to the situation and using only chemicals. None of the labels on the bottles ever suggested it was anything other than the real thing. No doubt those involved considered themselves creative problem-solvers rather than criminals – sometimes it's a matter of perspective.

Not unexpectedly, the desire to earn a little extra cash is most rampant during tough economic times. Yet these are also the times when people may be most tolerant of food fraud. Low-income households are looking for cheap calories and may turn a blind eye to even the most questionable of deals. In the mid-nineteenth century, putrid meat and cheese would be 'polished' by adding a layer of fresh fat or cheese to the surface to make it appear edible. Fishmongers would paint the gills of decaying fish to make them appear fresh. The vendors got away with this because the poor working class would do their shopping on a Saturday night after they got paid and it was easier to disguise foul products in the low light conditions of the evening market. People keen to appear wealthier than they were would buy dodgy alum-laced loaves to impress guests with their fashionably white bread. Yet they knew, based on the price, that the deal was too good to be true. Between trying to keep up with the Joneses and simply providing for their families, those living on limited incomes in the 1800s often knew they were being swindled, but had limited choice. Today people with limited incomes are still the most vulnerable to fraud. Highly processed foods are often a cheaper alternative to fresh fruit and vegetables; there are not only more opportunities to adulterate a processed food, but there is also a lower risk of being caught. Perhaps the worst fraud of all, however, is that many of these foods contain empty calories – they are high-calorie food with little nutritional benefit. But this is perhaps a different story.

While economic gain sits on one side of the fraud scale, risk sits on the other. Low inspection rates reduce the risk of fraudulent food being detected, particularly when it comes to imports. The FDA, for example, has increased its overseas inspections, but still only inspects a very small fraction of the facilities that supply food to the US. Approximately 1 to 2 per cent of imports are physically inspected and only about 0.5 per cent will be tested in a lab. When you consider that the US imports 91 per cent of its seafood, this presents a reasonably low risk for fraudsters exporting to the US. Undeniably this is, as Kircher stated, looking for a needle in a haystack. With limited resources for inspections, priority will always go to checking foods that are known to pose a medium to high risk to public health – *i.e.* food with a history of causing foodborne illnesses. Under this model, testing for fraud will remain a reactive process.

Yet increasing the risk of detection can be a powerful deterrent for fraudsters. In 1995, the UK's Ministry of Agriculture, Fisheries and Food (MAFF) – now known as the Department for Environment, Food and Rural Affairs (Defra) – reported that 35 per cent of commercial maize oil collected off grocery store shelves was adulterated with undeclared oil. Advancements in the detection methods used to test maize oil, which we will go into detail about in Chapter 3, increased the sensitivity of detection, and six years later when the tests were repeated, none of the oil samples from the shops had been adulterated to detectable levels. The risk of getting caught had increased and this alone was enough to deter the fraudsters without any changes to regulations.

As well as low risk, fraudsters are looking for low effort. There needs to be an opportunity to commit fraud and it needs to be relatively easy, though it's likely that even the laziest criminals are willing to put in some effort for the right amount of money. The long and largely anonymous food supply chains and plethora of consumer-ready products present countless opportunities. Our food whizzes about the globe at an astonishing rate. An extreme example of this was provided by Felicity Lawrence, a special correspondent for the *Guardian* and author of numerous books on food labelling. She tracked a plastic tray of stir-fry baby vegetables, purchased from Marks & Spencer for £2.99 (US$4.70). The tray included asparagus shoots, miniature corn, dwarf carrots and leeks, tied together with a single chive. She found that the chives, trays and packaging had all been shipped from England to Kenya. There, women working in refrigerated packing sheds next to the Nairobi airport carefully tied the Kenyan produce (the asparagus, corn, carrots and leeks) together with the English chives and placed them on the plastic trays, which were then wrapped in plastic and shipped back to the UK. The chives had made a 13,679-kilometre (8,500-mile) round trip, and in all honesty most likely ended up in the bin or the compost.

The problem is that rarely do our products these days come through a straight-up-and-down supply chain; more often than not, it is a supply network. A comforting mug of hot chocolate on a cold winter day has approximately 31 points in its supply network – between the sugar, milk, cocoa and packaging, 31 businesses have had a hand in the final product. Seen another way, there have been 31 opportunities for adulteration. It was the complexity of this supply network that made tracking the source of the horse meat in the UK so difficult. The 'beefburgers' sold at Tesco, for example, were supplied by a factory that was sourcing its ingredients from about 40 different suppliers. The mix of ingredients going into the burgers can change on the production line every half hour, making it even more challenging to pin suppliers to particular batches. Multiply this challenge by the number of ingredients in a ready-made meal like lasagne and suddenly the chain begins to look more like a web.

Just as increasing the risk of getting caught can tip the scales on fraud, so can reducing the opportunities. It is yet another reason, among many, to try and reduce the number of steps from farm to fork. The UK now imports approximately 50 per cent of all its food, whereas 30 years ago it imported just over 25 per cent. Campaigns such as

'Buy British' or the 100-mile diet are steps towards a sustainable food system that supports local producers, builds long-term relationships with suppliers and has potential health benefits as a result of eating seasonally and locally. But there are sustainability claims against buy-local campaigns. These argue that it is environmentally less costly to produce food in some areas of the world as the fewer additions required for growing (fertilisers or day-length) outweigh the additional travel costs. Importing peppers from Spain, for example, may have a smaller ecological footprint than growing them in polytunnels or greenhouses in Britain. However, an additional benefit of sourcing food more locally is that it removes some of the anonymity from the supply chain.

To give local producers a competitive edge over cheap imports, government tariffs are often introduced on imported goods. Yet these tariffs can, unfortunately, be a trigger for food fraud. In 2003, the US International Trade Committee (ITC) agreed with US catfish farmers that the importation of Vietnamese catfish was causing losses to the US market. US catfish farmers simply couldn't compete with the price of the Vietnamese imports in what became known as the catfish wars. The ITC imposed higher tariffs – raising catfish import duties from 5 per cent to upwards of 64 per cent. In theory, this should have evened out the playing field for the US farmers and encouraged the purchase of US products. An unintentional side effect was that distributors began importing Asian catfish as grouper – a higher-value finfish – to avoid the tariffs. One import company alone avoided approximately US$63 million in tariffs; the ex-CEO of Sterling Seafood, a New Jersey seafood importing company, admitted to importing more than five million kilos (11 million pounds) of Vietnamese catfish between 2004 and 2006 and selling them under the labels of grouper and sole. When the catfish–grouper scandal started breaking in 2007, grouper consumers lost their confidence in the product. Subsequent studies showed that more than half of consumers changed their seafood purchasing habits after this scandal broke, turning to non-grouper products instead. This no doubt had a significant impact on places like Florida where commercial landings of grouper were worth about US\$21 million (£13.4 million) at the dockside in 2007.[4]

While fraudsters need to have the opportunity to commit fraud, it also needs to be relatively easy. Though there are numerous examples of sophisticated fraud (the fake eggs we mention in the Introduction for one), the truth is that the easier it is, the more likely it is. Mixtures and ground-up foods, such as cheap oil mixed with expensive oil, papaya seeds ground with peppercorns and horse minced with beef, are at the opposite end of the spectrum to making elaborate fake eggs. For these easier to adulterate foods, there needs to be either less opportunity or greater risk of getting caught in order to tip the scale on the fraudsters.

Uncertain futures = more food fraud

While the economic rewards and the opportunities to commit food fraud are unlikely to disappear any time soon, there is yet another looming reason to believe food fraud will be a part of our future. An uncertain environment, brought about by climate change, is likely to generate the conditions that can trigger fraudulent acts. Predictions are that we will see an increase in extreme weather events, such as flooding and drought, which can have serious implications for agricultural yields. We've had many examples of this on a small scale already. In 2003, the UK experienced exceptionally hot weather and most UK lettuce matured quickly and all at once. Producers had a commitment to supply supermarkets with lettuce throughout the summer, but with UK lettuce finished, they had to import lettuce from the US to fill their orders. They sold the lettuce at a loss in order to maintain good relationships with the supermarkets. The agreements between most supermarkets and suppliers put the onus of a fluctuating demand or supply on the suppliers. A supplier must provide a certain quantity of product to the supermarket, but if the supermarket sells less than expected, they are not obliged to take the whole quantity. This leaves the supplier with an unexpected surplus they have to unload or waste. Yet, alternatively, if there is more demand than the supermarket expected, suppliers have to source the product from elsewhere. More extreme weather in the future could compromise the stability of the supply chain, putting suppliers in an even more challenging position.

Extreme weather events can lead to economic instability, disruptions to the infrastructure we depend on for our food supply and many other trickle-down effects. Let's consider rice for a moment, a staple food for half the world. The global rice market was estimated to be worth £14.6 billion (US$23 billion) in 2011. Owing to its very specific growing requirements, rice production is concentrated in South and South-east Asia. Thailand is the world's largest rice exporter, producing 28 per cent of global rice exports; Vietnam and India round out the top three rice exporters. These are three countries that are extremely vulnerable to the effects of climate change, such as rising sea levels and extreme weather events. We have a commodity that is both concentrated in terms of where it can be produced and vulnerable to climate change. The global supply of rice could be significantly affected in the future with few alternative growing regions.

China is the world's biggest consumer of rice, munching through approximately 127 million tonnes of the fluffy grain each year. The country has historically been self-sustaining in its rice needs as it is also the world's biggest producer of rice. Since 2011, however, China has increased its imports fourfold. This increased reliance on other countries for their favourite grain is partly due to rising consumption, but also because drought has hurt China's crop production, sending domestic prices higher. Future climate change could perpetuate this problem, increasing China's reliance on imports. So, what happens when a country with a voracious appetite for 127 million tonnes of rice enters a global market that trades a comparably measly 33.5 million tonnes of rice each year? We could see markets destabilise and prices could soar in response to a global shortage of this major staple food for some of the world's most impoverished people. This could be a red flag for food fraud; we have seen rice scams in the past, including fake rice made from plastic resin and potato starch.

Climate change may also increase the prevalence of disease. Milder weather in south-west England in the last few years, for example, has increased the infectious period of some livestock diseases, such as blowfly strike. Adult flies are emerging earlier in the spring and infecting animals for longer. With 80 per cent of UK sheep flocks affected by these parasites each year, an increased prevalence of the disease can have serious implications not only for animal welfare, but also for industry yields. Low yields due to extreme weather and disease, economic hardship brought on by natural disasters and interrupted movement of food globally – these are all the conditions that could force otherwise honest people into shady territory.

The effects of climate change will most certainly not be limited to the land. Over the last 25 years, temperatures in the North and Baltic Seas have risen five to six times faster than the global average. As a result, there has been a shift in the species found there. Cold-water-loving species, such as cod, haddock and whiting, have declined in numbers while warm-water-loving species, such as red mullet, red gurnard and John Dory, have all increased. Even as little as a 1–2°C change in water temperature around the UK could mean very different species living in these waters. This could present a problem for British consumers, who tend to be very traditional in their seafood choices. Approximately 80 per cent of seafood bought in UK supermarkets is cod, haddock, tuna, salmon or prawns. Four of these five species are either farmed (salmon) or imported (cod, tuna and prawns). There have been large pushes from supermarkets, governments, non-government organisations (NGOs) and celebrity chefs to educate the British public about alternative species, and to make people more comfortable cooking and eating UK-caught species such as squid, sardines, coley and dab. While these campaigns to eat lesser-known fish can be very successful in the short term, Brits generally go back to their favourite five over the longer term. This inflexibility among British consumers to try new species may increase their vulnerability to mislabelling and fraud in the future. There is an economic incentive to label a species as a fish that's in demand and there is more opportunity to mislabel when seafood is imported.

So, with greed, economic hardship, disasters and globalisation of our food supply all fuelling food fraud, we should expect to continue being swindled in the future. The question is whether we are likely to see more of it or whether we can start to tip the scales on fraudsters. Improvements in detection technologies and forecast systems, such as the one being developed by the NCFPD, could increase the risk of detection. Industry changes that improve the integrity of the food supply network and changes to consumer behaviour could reduce the opportunities for fraud. Will it all be enough, though, to make us more confident consumers in the future?

The cheap food conundrum

Perhaps before we can become confident in what we buy, we need to be realistic about our food. While globalisation of the food supply opens up new opportunities for sourcing produce and ingredients, it has also created unrealistic consumers. We now have seemingly unlimited choices at the supermarket. We can buy fresh lettuce in December just as easily as we can in July. We have, for the most part, lost touch with eating seasonally, we have become nothing short of delusional in terms of what food should cost and there are many steps between the food we buy and the people who grow it.

In 1900 the average family in the US spent 43 per cent of their income on food, while in 2013 this had decreased to 13 per cent. Conventional food systems have maximised efficiency to provide cheap food and consumers have come to expect bargain buys. Supermarkets keep the price of certain staples and highly processed foods low to remain competitive. For example, in 2012, scorching heat and severe drought affected over 80 per cent of the US corn crop – a staple of animal feed and processed food ingredients. That summer the US government predicted the low corn yield would cause a 4 to 5 per cent increase in the cost of fresh produce such as beef, pork, eggs and dairy – well above the average inflation increase of about 2 to 3 per cent. Yet US supermarkets managed to keep the cost of packaged food products low. How is this possible when surely the cost of the ingredients going into those packaged products increased?

The highly processed packaged food industry is very efficient, which is why it costs more to buy a bag of apples than a box of macaroni cheese. It is also far more expensive and labour-intensive to transport fresh fruit and vegetables to the supermarket without spoiling than it is most prepackaged food. Between cost and convenience to the customer, it is not surprising that ready-made meals are one of the fastest growing sectors in the food industry. In the UK, high-end ready-made meal sales in supermarkets topped £2.3 billion (US$3.6 billion) in 2013 – a growth of 8 per cent over the previous year. The growth was attributed to consumers being more cash-conscious and switching from a meal out to a ready-made meal in. In 2007, the US imported US$60 billion (£38.1 billion) worth of consumer-ready food, including processed food products, which was a 100 per cent increase over 1998 consumer-ready imports. Despite more than half of consumers reporting that they plan to eat fewer prepared meals and cook more, purchasing behaviour would suggest otherwise.

It is possible, however, that the days of finding a good deal are nearing an end. Many believe that the cheap food era is over. In 2013, the United Nations (UN) predicted a 40 per cent rise in global food prices over the next 10 years. They credit a growing middle class in countries such as China and India with an unprecedented demand for meat. If the cheap food era is indeed over, we as consumers may be particularly vulnerable to fraud during this time of transition. There could be a substantial gap between what deal-seeking consumers are willing to pay for food and the actual cost of production – the perfect conditions for tempting fraudsters to source cheaper ingredients, such as horse meat.

Horse kicks governments into gear

While the horse meat scandal luckily did not compromise anyone's health (in fact it may have actually improved it owing to the lower fat content of horse compared with beef), it did help awaken us all to our vulnerability when it comes to food. At some level, the whole thing seemed like some schoolboy prank. As though somewhere a group of bullies were laughing and pointing fingers and saying 'You'll NEVER guess what we put in little Tommy's burger!' As the saying goes, fool me once, shame on you; fool me twice, shame on me. The horse meat scandal prompted governments around the world to avoid being made a fool of twice.

Since Horsegate, food fraud seems to have moved up government priority lists. Britain launched a review into its food supply networks, led by Professor Chris Elliott from Queen's University, Belfast, and subsequently introduced a dedicated Food Crime Unit. In late 2013, the US FDA released a new proposed rule to protect food from intentional adulteration. The European Union (EU), previously more focused on food safety issues, has now mapped the tools and mechanisms that exist within their member states for fighting food fraud. The EU is also in the midst of assessing and developing EU-wide protocols for testing for fraud and they are in the process of extending mandatory origin labelling to more food products, including unprocessed meats (such as horse), milk and single-ingredient foods. The general trend is that governments are trying to be more proactive on food fraud, either by increasing the risk of detection or by reducing the opportunities to commit it.

There also seems to be a push, at least within the UK, to shift enforcement bodies from a very administrative, ticking-the-right-box approach to a more policing and investigative approach. Professor Elliott, in his investigations, came across one meat processor that had been audited 300 times in the previous calendar year and the biggest issue to crop up was that the fire extinguishers needed to be moved. Inspectors might need to start channelling their inner Sherlock and move beyond placement of fire extinguishers (still important) to opening closed doors leading into back rooms and closets, scrutinising the books and generally investigating things that simply don't look right.

Some countries have fully embraced this policing approach to food inspections. In 2006, the Danish Food Administration formed the Danish Food Flying Squad. Journalists had uncovered fraud within the meat trade in Denmark – tonnes of expired meat was being circulated in the market. The Squad started with six people and by 2013 had grown to 18. Just as its name suggests, the Flying Squad arrives at a location by helicopter, unannounced, and the business has no more than five minutes to prepare for the inspection. The inspectors have a no-closed-door policy so if doors are locked and keys are conveniently misplaced, a locksmith is called immediately. They have a culture of suspicion because they believe that if you don't look specifically for fraud, you won't see it. The Flying Squad conducts routine inspections, but also follows up on consumer complaints and anonymous tips. The Flying

Squad inspectors have more authority than the police in terms of entering premises and they have the authority to confiscate equipment, collect and document evidence and hand out warnings and fines. The UK's Food Crime Unit has already said it will be taking some lessons from Denmark in terms of its approach.

Can we blame government?

If governments hadn't started to take action in the face of such gastro-affronts as horse meat in burgers, there would have been a public outcry. When the horse meat scandal hit Europe in 2013 consumers were blaming retailers, retailers were blaming suppliers and everyone was blaming the government. Which begs the question of who is really responsible? Who can we rely on to differentiate the beef from the bull?

As everyone likes to blame government, let's start there. A government's role is to develop the legal framework that secures the rights and freedoms of its citizens. As well as ensuring that certain rights and safeguards are in place, the government must also then enforce them.

In the UK, enforcement of food safety and food standard controls on food products is the responsibility of local authorities – with the exception of certain meat facilities, which are the responsibility of the FSA. Food quality, hygiene and safety issues are dealt with by environmental health officers in district councils. Composition and labelling of food products is enforced by trading standards officers in county councils. Many of these local authorities rely on contractors to help them complete their inspection programmes as they simply don't have the capacity to fulfil all of their duties. Since 2008, there has been a reduction in food law enforcement staff in 63 per cent of local authorities.[5] Local authorities are working with limited resources and have the discretion to set their own budget priorities; food law enforcement may not be top of the list.

The local authorities work with and are audited by the FSA. This is the agency in the UK tasked with protecting public health and consumer interests in relation to food and drink. As well as working with local authorities on enforcement, the FSA works with businesses to make sure they are delivering safe food products. This includes safety-related food labelling, such as allergens and 'use by' dates.

While the FSA is responsible for safety-related labels, there are two other government departments in England with responsibilities around food labelling as of 2010. Nutrition labelling is the responsibility of the Department of Health, and food authenticity and composition labelling is the responsibility of Defra. The National Audit Office attributed this montage of government responsibilities as one of the contributing factors that enabled the horse meat scandal to occur. There is confusion among stakeholders as to who to contact or where to get information and it weakened the government's ability to share intelligence. This distribution of responsibilities was also one of the first realisations made by Professor Chris Elliott in his review of UK food supply networks. He claimed that although each of the government departments works hard in its area of responsibility, they aren't working together. In response, the UK government set up the Food Integrity Committee with members at the ministerial level to help gather up the fragments of food responsibility within government.

The UK government isn't alone in this division of responsibility around food. The FDA and the US Department of Agriculture (USDA) take leading responsibility for food safety in the United States. The FDA is responsible for the safety of all domestic and imported food products except for meat and poultry, which land on the plate of the USDA. Eggs in their shell are the responsibility of the FDA, but crack them open and they become the responsibility of the USDA's Food Safety and Inspection Service (FSIS). In 2013, the Canadian government moved responsibility for the Canadian Food Inspection Agency (CFIA) over from

Agriculture to Health Canada. Though there are still three bodies with responsibilities for food – CFIA, the Public Health Agency of Canada and Health Canada – they are at least all under the responsibility of one minister. In China, food safety is the responsibility of no fewer than 10 different government departments.

In terms of food fraud, governments are also well positioned to collect, collate and share information. The Netherlands, with their long history of food trade and therefore an equally long history of food fraudsters, have extensive experience in gathering information on food fraud. In 2013, the Netherlands Food and Consumer Product Safety Authority (NVWA) gathered 120,000 pieces of information around food fraud. All of that information turned into just 20 actionable cases. The US Pharmacopeial Convention (USP) hosts the Food Fraud Database, which provides details on foods that have been reported as adulterated in the media and peer-reviewed literature. It states the adulterants found in the food and the detection methods used to find them. Searching through the database gives you a very basic indication of what foods are most vulnerable to fraud and the diverse adulterants that have been used to stretch them out. After Horsegate, the EU realised that no statistics existed on the incidence of food fraud in the EU. Gathering information is the first and most basic step in tackling the issue; it builds an understanding of where the vulnerabilities lie and eventually can support actions. In addition, it's critical to share this information because just as food moves around the globe and criminals operate across different countries, information must be equally mobile.

Government also has a responsibility to deter criminal activity through law enforcement. While there seem to be endless examples of food fraud, prosecutions related to food fraud are far more elusive. A year after the horse meat scandal in the UK came to light, there had still been no prosecutions. The first prosecution in the UK related to horse meat sales (not Horsegate itself) did happen late in 2013. A random test on pork sausages bought in Dartford, England exposed its equine side; it contained nearly 50 per cent undeclared horse meat. The food importer, Expo Foods Ltd, was fined £5,000 and a further £2,500 (combined, around US$11,400) to offset the council's investigation costs. The first charges directly related to Horsegate weren't laid until March 2014, 15 months after the discovery of the undeclared horse DNA in frozen burgers.

Despite the existence of the legal framework for prosecuting food fraudsters, there seems to be a gap between the possible penalties and the punishment handed out to these criminals. In 2012, two executives of the Hitchin' Post Steak Co., a poultry-slaughtering and processing business in the US, were indicted on charges of selling misbranded and adulterated poultry products. Their alleged crimes carried the following *maximum* penalties upon conviction:

- Five years in a federal prison and a fine of US$250,000 (£158,000) for conspiracy to transport and sell misbranded or adulterated poultry.
- One year in a federal prison and a fine of US$100,000 (£63,500) for unauthorised use of an official mark of inspection.
- One year in a federal prison and a fine of US$100,000 (£63,500) for selling or processing poultry products that were misrepresented as having been inspected.
- Three years in prison and a fine of US$250,000 (£158,000) for selling adulterated poultry products.
- Three years in prison and a fine of US$250,000 (£158,000) for unauthorised use of an official mark of inspection on adulterated poultry products.
- Three years in prison and a fine of US$250,000 (£158,000) for misrepresenting that the poultry products had been inspected.

If they had been convicted and sentenced to maximum penalties this adds up to 16 years' imprisonment and a potential US$1,200,000 (£762,700) in fines. What actually happened was rather different. Both of the Hitchin' Post executives entered plea agreements; the Vice President was sentenced to one year on probation and a US$5,000 (£3,178) fine, and the General Manager was sentenced to one year on probation and ordered to pay US$15,453 (£9,822) in fines and restitution. With minimal penalties, food can be a profitable arena for organised criminal activity.

Following the melamine milk scandal in China, 21 people were tried for their involvement. The head of Sanlu was sentenced to life in prison and fined more than US$3.1 million (£2 million). Farmer Zhang Yujun, who produced and sold hundreds of tonnes of melamine-laced protein powder, was executed in November 2009 along with middleman Geng Jinping, who sold more than 900 tonnes of tainted milk.

There is clearly great disparity in terms of the penalties associated with food fraud. It would seem logical, however, that if we want to tip the scales on fraudsters, the punishment must at least be enough to counter any financial benefits of committing the crime. Heavier jail sentences and fines for those who are caught might also make a more convincing deterrent. In November 2014, Taiwan boosted fines for food fraud 10-fold to try and restore confidence in their consumers after olive oil made by Wei Chuan Foods Corp was found to contain restricted food colouring.

Professor Elliott identified zero tolerance as one of the pillars of food integrity. When he described this pillar at a conference he made reference to a concept in criminology known as the 'broken windows theory'. Very simply, the theory is that if a building has a few broken windows that go unrepaired, vandals will break a few more windows. Fix those windows quickly, however, and vandals are less likely to break them. The bottom line is: fix the problems even when they're small, and punish even small fraudulent acts with stiff penalties. Enforcement is where government can take a leadership role.

The Danish Flying Squad handed out approximately half a million pounds' worth of fines in the first year of its operation. Seven years later, it is handing out about a quarter of these fines and this is largely to do with increased compliance. Businesses don't want to get caught so they have improved their self-regulation. Not only does the Flying Squad hand out fines and then bill the business for their time, they also name and shame the companies by publishing the names of businesses found breaking the law and issuing press releases to get the word out. Beyond that, they've even had a TV crew follow them on the job – much like the reality show *Traffic Cops* or the US equivalent *Cops*.

The responsibility of industry

Tough penalties will motivate the food industry to do a thorough audit of their supply networks because they do not want to be held responsible for fraud. It goes without saying that the food industry – from producers to packagers to distributors to retailers – has responsibility when it comes to food fraud. The industry bears considerable economic cost when fraud is committed – recalled products, increased testing and lost sales due to lack of consumer confidence. It therefore has a vested interest in ensuring the integrity of the supply network, knowing where the areas of risk are in the supply lines and how the business would respond in the event that the supply network is compromised.

KPMG, a global network of audit, tax and advisory professionals, works with food industry clients to examine and manage their supply risk. As part of the analysis, clients map their supply networks from end to end, looking at the multiple levels of their suppliers (*i.e.* who supplies their suppliers). They consider all the risks they are exposed to through their supply network and how a global event could have an impact on their supply. They take a close look at their auditing processes – their own as well as those of their suppliers. This is no small task. Walk up and down the aisles of a modern supermarket. Think about the number of products on those shelves and then the number of ingredients in each product. The task of resolving where they come from, let alone ensuring their integrity, seems insurmountable. The number of products stocked by the average US supermarket went from just under 9,000 in 1975 to 47,000 in 2008. The average supermarket in the UK stocks a relatively sparse 30,000 items. Much of this growth stems from new varieties of the same product – there's no longer one box of Cheerios on the shelf, but 11 varieties of Cheerios. While some variety is good, imagine how this increases the supply network. Original Cheerios has seven ingredients; with the expanded range, the ingredient list more than triples. The supply network has virtually exploded, but the systems in place to manage that network have not kept pace with this expansion.

As confirmed by the Elliott report, one of the biggest changes industry can make towards reducing its risk of exposure to food fraud is to shorten its supply chains. One of the retailers held up as a good example of this is McDonald's in the UK. Perhaps not a name you were expecting? In the UK, beefburgers have three stops before they reach McDonald's – farm to slaughterhouse to processor. The restaurant chain buys all its beefburgers from the processor OSI Food Solutions and has done so since 1978. Linden Foods, a slaughterhouse and boning plant, has supplied OSI for 15 years. Linden Foods can trace each batch of meat back to the farm it came from. All shipments between locations are labelled and sealed to avoid any tampering and electronic tracking ensures that the number of patties going into restaurants equals the number going out. The supply chain is short and the relationship with the suppliers is long term.

The UK government has stated that it will support industry's efforts to put in place a robust and effective supply chain audit system. NSF International, independent experts in food safety and quality, are working with the FSA to develop tools for industry that help them identify products that are vulnerable to fraud.

Alongside the top-down support from government, industry could be further motivated to make changes with bottom-up support from its customers. At the moment, however, there seems to be little evidence that the integrity of the food chain or fraud more generally is a primary consideration for most consumers. We consumers are complex creatures – a behaviouralist's nightmare. What we buy depends on who we're with, our mood, who we're buying the food for, recent trends and news headlines, our current account balance and even the weather. However, the recurring drivers in our purchasing decisions over the long term are convenience, health, enjoyment and value for money. At the moment, there's no proof that retailers and restaurants will gain any competitive advantage for the integrity of their food supply network. Yet, Waitrose – known for its commitment to small local producers, rigorous sourcing policies and sustainability – came out of the horse meat scandal unscathed and reported an 11 per cent boost in sales in its wake. This was likely to be as a result of discouraged Tesco shoppers moving over to Waitrose. In 2014, Waitrose experienced the biggest period of expansion in the company's history, while Tesco reported a 6 per cent fall in profit, and then later reported that even these profits had been grossly overstated. As technology becomes available that enables consumers to have instant access to information regarding the supply chain, retailers that have that information may have a distinct advantage over others.

To combat fraud, industry will also need to share information with government (and presumably vice versa). If governments are to be repositories of information, which ultimately leads to the identification and prosecution of criminals within the system, then they need information.

Andrew Opie, Director of Food and Sustainability for the British Retail Consortium (BRC), says the retail sector has started to do this. Not only have retailers carried out an extensive review of how they share information, but they've also looked at ways to shorten supply chains, and the BRC has developed new tools for auditing the supply chain that directly target food fraud. Things appear to be heading in the right direction.

Watchdog organisations and watchful consumers

Independent organisations operating outside of both industry and government, such as Oceana and Which?, also have an important role to play in combating food fraud. Many organisations are conducting their own targeted testing of products. Which? has helped to break the news of numerous food fraud scandals in the UK. They operate by conducting research and testing, which generates media attention. This in turn gets the public activated about the issue the organisation is campaigning on, thereby fuelling their work to change policy.

The media also have a role in the fight against food fraud. They must find the appropriate balance of ensuring that food fraud is kept in the consciousness of the public without overwhelming them with so many stories of gloom and doom that people are paralysed into a state of apathy. Beyond just reporting scandals, the media have also been critical investigators in breaking them. Felicity Lawrence and her colleagues at the *Guardian* have conducted numerous in-depth investigations involving the food industry – from undercover footage of unhygienic chicken processing to the exploitation of immigrant workers. It was Danish journalists who broke the expired meat scandals that led to the establishment of the Flying Squad. Back in the nineteenth century, the esteemed publication the *Lancet* printed weekly articles about the adulterated food found around London and publicly named and shamed the retailers selling these products – address details and all.

Media stories, if reported well, play a critical role in educating the public, though they rarely offer consumers suggestions on how to reduce their vulnerability to fraud. They do a splendid job, however, of keeping pressure on government as well as industry. The media could even help to build an environment where integrity of food products provides a competitive advantage within the industry.

We as consumers must also take some responsibility for food fraud. As already mentioned, we tend to have unrealistic expectations of what food should cost, but we've also lost touch with how food should look, smell, taste and behave. Saffron threads should look and smell a particular way, fresh olive oil tastes very different from old olive oil and pure freshly squeezed orange juice should not last for a week in the fridge.

There is such a plethora of intense synthesised flavours in our food these days that it may be affecting our food preferences. A child that has been exposed to artificial strawberry flavour in ice cream, sweets, gum, milk, cereal and goodness knows what else all of her life can still perceive the taste of a real strawberry. Yet she will probably have a preference for the fake strawberry flavour she has been raised with. After all, it's far more intense than a real strawberry, the flavour is consistent and it lasts longer on the palate. But what does this mean for our understanding of what natural foods should taste like?

If you've bought or made proper fresh bread, you know that it's completely unnatural for a loaf to last a whole week without becoming stale or mouldy. Commercial bulk-made loaves include preservatives and are dipped in antifungals in order to keep them fresh. This obviously isn't fraud. It's an important process that keeps food from being wasted. It is, however, an example of how we have become accustomed to how our food is processed without truly considering what that means.

We have a very sophisticated built-in fraud detection system. There is a common misconception that humans have a poor sense of smell. We have relatively small snouts compared with other animals and fewer chemical receptors. Yet behavioural tests have shown that we are equally good, if not better, at smell perception than some of our fellow mammals. We even outperform canines for some odorants. Trichloroanisole (TCA; see Appendix), which is a natural compound that can give wines a musty or corky flavour, can be detected by some people down to a few parts per trillion. So the system is in place, we just need some training in how to use this rather sophisticated equipment – a fine-tuning if you will.

All of us, from government to consumer, share in the responsibility for combating fraud in our food supply. But nobody should be blamed for food fraud except the criminals themselves. The best thing we can do as consumers is arm ourselves with some knowledge and, where available, tools to help reduce our vulnerability to fraud. Professor Lisa Jack, head of the Food Fraud Group at the University of Portsmouth, has given an analogy of having a police station in your community. There's clearly a police presence, but you still lock your door when you leave the house and take precautions against theft. The same should be true of food fraud.

And then there's science …
Science has an incredible role to play in all of this. There are many cutting-edge techniques that are helping to detect fraud – from DNA analyses to identify your fish fillets, to stable isotope analyses that can trace the origins of a tomato back to a sun-kissed valley in Italy. However, just as the detection techniques are getting more sophisticated, so are the scientific methods for committing fraud. If the fraudsters know that watered-down milk is going to fail a test owing to its low protein content, they find a way to adulterate it further to get it past that test (we'll talk about this in Chapter 6).

These criminals are getting creative in how they mess with our most basic ingredients. Who would think that a grain of rice would be anything other than the seed of a plant? Sure, we might expect some mislabelling of basmati, but to actually make a fake grain of rice by mixing together a combination of potato starches and plastic resins? What about building an entire egg, shell and all? Well, it's disturbing, yet there's also something rather incredible about it.

We wanted to write this book to share some of the outstanding science behind food fraud – detecting it as well as committing it. Partly we have felt that in all the coverage of food scandals, the science has been largely overlooked and, being science geeks ourselves, we think this is too interesting to ignore. But also we feel that knowledge is a powerful tool and understanding what's possible makes for more informed consumers. Science can give consumers the tools to help them reduce their vulnerability – to lock that door against food fraud. From kitchen chemistry to new technologies in the pipeline, there are things we can do to sort the beef from the bull.

CHAPTER TWO

Busting the Food Cheats

It was a German chemist living in London in 1820 who launched the first scientific attack on food fraud. Frederick Accum was not only a chemist, but also a great lover of food. He was distraught about the state of food in London at the time and began a one-man crusade to expose it for what it was. At the heart of his endeavours was his landmark book *A Treatise on Adulterations of Food, and Culinary Poisons*. Within its 360 pages lie numerous shocking accounts of food fraud, which opened the eyes of British consumers to the many ways they were being deceived in what they were eating and drinking. Accum's book offered a number of easy methods – almost kitchen chemistry – for detecting some of the most common adulterations of food. He described a method for detecting alum (aluminium sulphate) in bread, using boiled distilled water, paper and pure nitric acid (no doubt a rare staple in most kitchens). There is great detail on how to recognise spurious tea leaves and test them using ammonia, a common ingredient in household cleaners. There were tests for counterfeit coffee, adulterated wine and other alcoholic beverages, cheese, spices, pickles, vinegar, cream, confectionery, custard, olive oil and mustard. There were tests to reveal sinister, even toxic or deadly materials such as red oxide and other colourants. Accum's discoveries were a revolting revolution for the citizens of London.

Accum's *Treatise* has many dispiritingly similar resonances with the questions asked of today's food fraud detectives. The table of contents lists all too familiar vulnerable candidates for adulteration: wine, beer, spirits, tea, coffee, water, bread, dairy products, vinegar, spices, olive oil, pickles, fish sauce, *etc.* And the line in the preface to the *Treatise* 'To such perfection of ingenuity has the system of counterfeiting and adulterating various commodities ... found in the market, [been] made up so skilfully, as to elude the discrimination of the most experienced judges' remains as true today as it was 200 years ago. Accum's tests were a reflection of the embryonic state of analytical science at the time; nevertheless, they were able to reveal what we would now regard as crude, almost desperate deceptions. For the most part, his tests involved simple solubility, combustion, weighing, distillation and chemical 'spot' tests, requiring no sophisticated instruments. However, Accum's work was just the beginning.

Arthur Hill Hassall, a British physician, chemist and microscopist, picked up on problems identified by Accum and expanded upon them using his microscope. Between 1851 and 1854, Hassall put over 2,500 food samples under microscopic scrutiny and published his findings (good and bad) in the *Lancet*. Hassall was highly influential in bringing into British law the Food Adulteration Act in 1860. This was reinforced in 1872 by the Adulteration of Food and Drugs Act, which was supported by the establishment of public analysts and associated enforcement officers. In 1874 the Society of Public Analysts was founded and Hassall served as its first president. His work within the Society would eventually help form the Sale of Food and Drugs Act of 1875, which introduced heavy penalties for the adulteration of food.

Around the same time, across the pond, Lewis Caleb Beck, an American physician and professor of botany and chemistry, was beginning chemical analyses of agricultural products. In 1848 Beck published his book *Adulterations of Various Substances Used in Medicine and the Arts*, and just like Accum, he provided the tests that could help expose these adulterations. Though Beck's investigations weren't directly of food, he is credited with helping to lay the groundwork for the Pure Food and Drug Act to be passed in 1906. It was

Harvey Washington Wiley, Chief Chemist of the Department of Chemistry within the USDA (United States Department of Agriculture), who was a major driving force behind the establishment of the Act, which offered radically new protection mechanisms for US consumers. Wiley published a series on *Foods and Food Adulterants* and later a textbook entitled *Foods and Their Adulterations*. Most famously, Wiley established the Poison Squad (given its name by a journalist) in 1902. It was a group of volunteers Wiley co-opted to measure the deleterious effects of various food preservatives, such as borax (chemically known as sodium borate) and formaldehyde (also known as formalin – the solution used in preserving biological specimens and for embalming human bodies). All of these studies helped form the technical frameworks for the first US food laws. The Federal Meat Inspection Act was passed in the same year (1906) as the Pure Food and Drug Act, and it not only ensured the safe and sanitary processing of meat and meat products, but also helped to prevent adulteration and misbranding of meat. The Pure Food and Drug Act converted Wiley's department to the Bureau of Chemistry and put it in charge of administering the Act. It was the Bureau of Chemistry that would later go on to become the US FDA in 1930.

Oh Dr Accum, how things have changed

Food fraud today is a global phenomenon – although the truth is that the actual scale of the problem is completely unknown. This is because no matter how sophisticated the tests become, it is impossible to screen every food item for its authenticity. The entire food industry is therefore faced with a major dilemma – a dilemma which becomes the fraudsters' opportunity.

While the motivations for food fraudsters have changed very little since Accum's time, the scale of the deception has moved into another league. Twenty-first-century food fraud detection now demands the analytical tools of twenty-first-century CSI-style forensic science. The tests of Accum's time, using materials which were more akin to those now found in a high school student's chemistry kit, are virtually useless. The fraudsters appear to be fully aware of the sophistication of the food forensic tests available and continually tweak their activities accordingly to beat the tests. For example, fertiliser or melamine (both high in nitrogen) are added to fake milk to deceive the simple tests for protein based on total nitrogen content. The economic gains of global food fraud are such that we are now all reliant upon an arms race in which the regulators are continually forced to refine their approaches in response to increasingly diverse fraudulent activities.

The way we purchase our food and the criteria we use to select it have also changed from Accum's time. In terms of fraud, fresh fruit and vegetables offer a degree of reassurance since you can plainly see what they are – an apple looks like an apple and a carrot looks like a carrot. The criteria of smell, taste, appearance and freshness are fundamental when selecting such foods, but all too often in modern supermarkets this information is masked by packaging – broccoli sealed in plastic, and apples and carrots in bags. The modern consumer also uses a wide range of other criteria, such as country of origin, mode of production and, of increasing importance, nutritional value to make purchasing decisions. For this information we have to rely entirely on labelling and certifications. Apart from clearly recognisable wholefoods, such as vegetables, fruit, grains, pulses, whole spices and whole animals, how can we be confident that what we are eating is what we actually want to eat? We place our total trust in the supplier, whether they be an individual market stall holder or a global supermarket chain. The supplier in turn places their trust in the producers, and this raises the problem of supply chains discussed in the previous chapter.

Food processing, while revolutionising food storage, has also introduced complexity into the food chains and into the foods themselves, which opens doors for fraudsters. Most foods are processed in some way; canning tomatoes, mincing meat, extracting seed oils and making butter are all examples of processing. Even a basic apple is processed; it's washed, graded, photographed to determine blush ratio, sorted and thrown behind a sensory deprivation force-field (plastic wrap or bag) ready for sale. We'd like to think that these simpler forms of processing are immune to fraudulent activities. Fresh tomatoes are clearly identifiable, but how confident can we be that canned tomatoes are purely tomatoes? How can you know you are not buying undeclared added water? Horsegate was proof that minced beef isn't always the all-beef product we're expecting. Butter looks and tastes like butter, but would you be able to detect a small proportion of added margarine or vegetable oil by taste or texture alone? And if it is supposed to be organic butter, can you be sure it was made from organic milk?

Remarkably, there are known incidences of US farmers losing their organic certifications because they unwittingly used a non-organic fertiliser that was being mis-sold as organic by leaving synthetic components (ammonium sulphate fertiliser) off the ingredients list. Interestingly, this was not picked up through analysis but by the enforcement authorities checking the fertiliser production plant. The supplier was jailed, so the laws definitely have teeth, but how can organic food be authenticated in the absence of such direct enforcement work? We will return to this in Chapter 9.

This is only the tip of the iceberg. There are even more opportunities for fraud among foods that are not as easily traced back to any plant or animal origins; that is to say those containing highly refined and even chemically produced ingredients, such as nature-identical substances, including chemically synthesised artificial colourants or flavourings. While we can read the ingredients lists on the sides of the packets, jars and bottles, few of us can fully understand what many of the ingredients actually are.

INGREDIENTS: WATER, SUGAR, CORN SYRUP, MILK PROTEIN CONCENTRATE, VEGETABLE OIL (CANOLA, HIGH OLEIC SUNFLOWER, CORN), COCOA PROCESSED WITH ALKALI, SOY PROTEIN ISOLATE, AND LESS THAN 0.5% OF POTASSIUM CITRATE, MAGNESIUM PHOSPHATE, POTASSIUM CHLORIDE, CELLULOSE GEL AND GUM, SALT, CALCIUM PHOSPHATE, CALCIUM CARBONATE, SODIUM ASCORBATE, SOY LECITHIN, CHOLINE BITARTRATE, ALPHA TOCOPHERYL ACETATE, ASCORBIC ACID, CARRAGEENAN, FERRIC PYROPHOSPHATE, NATURAL AND ARTIFICIAL FLAVOR, ZINC SULFATE, VITAMIN A PALMITATE, NIACINAMIDE, VITAMIN D_3, CALCIUM PANTOTHENATE, MANGANESE SULFATE, COPPER SULFATE, PYRIDOXINE HYDROCHLORIDE, THIAMINE HYDROCHLORIDE, BETA CAROTENE, RIBOFLAVIN, CHROMIUM CHLORIDE, FOLIC ACID, BIOTIN, POTASSIUM IODIDE, VITAMIN K_1, SODIUM SELENITE, SODIUM MOLYBDATE, VITAMIN B_{12}.

Figure 2.1. So this is food? Understanding the ingredients list on some food items requires an advanced chemistry degree in order to truly know what we're eating.

Deciphering these lists requires a high-level knowledge of chemistry, biochemistry and food technology, not to mention the capacity to translate the numerous additive codes into some incomprehensible chemical name. E385, for example, is calcium disodium ethylenediaminetetraacetate (EDTA), which is a preservative; E900 dimethylpolysiloxane is an antifoaming agent, which we'll meet again later. Even if you take the time to understand what all the ingredients actually are, how can you be sure that even these are what they say they are and they're present in the right amounts? Extensive blending offers numerous opportunities for introducing cheaper ingredients to improve profit margins.

Who's doing the testing?

The complexity of many commercial processed foods presents a unique challenge for the twenty-first-century human. Like all animals, we use our primary senses to assess the nature and quality of the food we eat based on appearance, taste, smell and texture. There's a reason our nose is strategically placed above where we put the food in! Food with added colourings, flavour enhancers, aromas and texture ingredients that improve the rheological or 'mouthfeel' properties of food make verifying its authenticity an impossible task for consumers.

One could quite reasonably expect that the authenticity of a given food item, processed or otherwise, must ultimately be the responsibility of the seller. But the array of foods on offer and the numerous ingredients used to prepare many processed foods makes their authentication by the seller a nearly impossible task. It is also relatively easy for sellers to lay blame elsewhere. Tesco, for example, was able to quickly lay the blame on its suppliers during the horse meat scandal as it had been assured the burgers being packaged under their Everyday Value label were all beef. But in the end, it was Tesco that suffered the loss of consumer trust. It is for this reason that many retailers are conducting their own authenticity testing. Since Horsegate, two UK retailers, the Co-operative and Sainsbury's, have confirmed the use of tests, such as stable isotope analysis (which we explain later in this chapter), to verify the origins of the meat they buy as part of their own authenticity testing programmes. However, they can't test the 30,000 or more products on their shelves, let alone the ingredients that go into them. Sellers are limited in what they can do.

Does the responsibility of authenticity testing then lie with the producers and suppliers? This is where the problem of complex supply chains comes to the fore. Even if a rogue ingredient can be detected, the complexity of food supply chains makes pinpointing its source a formidable challenge, especially given the huge numbers and volumes of products involved. In 2015, Reily Foods Company, a US food manufacturer, was conducting testing on its products in the wake of an adulterated cumin incident at the end of 2014. The company found that the cumin they were using in one of their chilli seasoning kits contained almond and peanut.

It triggered a second, larger scandal that would pull products off the shelves in the US, Canada and UK. These were mainly spice kits, but also croutons, soups and chilli products, which all contained cumin that had been adulterated with ground almond and peanut shells (we go into more detail in Chapter 7). When Reily Foods notified their US supplier of their test results, the FDA had to go to work because that supplier had also sold the cumin on to 38 other food manufacturing companies – this became a high-allergen risk incident. The US supplier had purchased the cumin from a Turkish supplier, but it hadn't been revealed at the time of writing this book whether the adulteration occurred in Turkey or whether the cumin was already adulterated when the supplier imported it. Though it may never be possible to pinpoint where the adulteration occurred, it was testing within the industry itself that helped pull potentially lethal products off the shelf in a reasonable timeframe.

Watchdog organisations, such as Which?, are doing testing, as are NGOs, but they will target specific food products related to their latest campaigns. Research is also being conducted by higher education institutions all over the world, as we will describe throughout the book. And while this research is critical in advancing testing methods, until recently there has not been a concerted effort in academia to tackle the confronting issues in detecting food fraud (though this may be changing). Government inspection agencies are also doing testing. After Horsegate, the FSA increased its food sampling programme budget from £1.6 million (US$2.5 million) to £2.2 million (US$3.5 million). But as we said in the previous chapter, these agencies are also responsible for foodborne pathogens and more acute food safety concerns, and resources are limited.

Despite the industry, retailers, NGOs, academics and government inspection agencies all doing testing, food fraud still happens. The stark reality is that the scale of food production is so massive that comprehensive routine screening is an impossible task. The best that can currently be achieved is random testing of what amounts to only a tiny fraction of all the food we consume. And this testing is conducted by a whole lot of different groups that don't necessarily share their results with each other. It is therefore inevitable that we will be regularly consuming foods that are not what they claim to be. This will especially be the case if our supermarket trolley includes a preponderance of high-value premium commodities or highly processed food. The fraudsters have, however, also realised that huge profits can be made from adding lower-cost ingredients to much more widely consumed staples. Misrepresenting popular specialist or beneficial commodities, such as organic foods or those of specific geographical origins, is another dimension of food fraud. Detecting such activities is a monumental task requiring the application of analytical approaches and techniques every bit as sophisticated as those forensic science methods used by crime scene investigators. But instead of a victim in a crime scene, food inspectors have a bottle of oil – and there may be absolutely nothing wrong with it. So where do you start?

Framing the food fraud question

Globalisation and commoditisation of our food has got us into a situation where we can't always trust what we eat. We have set our table, now we must sit and eat at it ... like it or not. And so we must regularly invite analytical science to dine with us as it is, without a doubt, one of our greatest allies in detecting food fraud. It is by applying the correct methods and approaches that we are able to confirm that the food sold to us is what it says it is. If we assume that the fraudsters are working in most areas of the food industry, then we need tests that will allow all commercial variants, even of the same commodity, to be scientifically distinguishable. Thus, the crucial first step is defining the problem. This involves recognising all the common types of a given foodstuff and understanding what the natural variation may be. Oil and wine are examples of this that we discuss later. There needs to be complete understanding of how the food is made in order to develop an awareness of the most likely ways it might be manipulated or misrepresented to the advantage of the fraudster. And from this list of potential manipulations, analytical scientists must choose which specific question to answer as there is no one test to answer them all. Tom Mueller's book *Extra Virginity* provides a case study for olive oil, which we will return to later.

This is the overarching framework within which food fraud detectives operate. Examples of the more specific questions they seek to answer are: is the honey pure? is the salmon farmed or wild-caught? is the minced meat pure beef? is the olive oil Italian in origin? is the butter organic? and so on. The more specifically we can frame the question, the more chance there will be of developing a test to answer it. These are extremely complex questions, but are typical of the challenges facing the detection of food fraud around the world on a daily basis.

An analytical test is more precisely referred to as a determination in the analytical profession. When it's applied, the result obtained, ideally a number, is compared with the accepted known outcome (numerical standard) for that food. We use such logic throughout our lives – we know what fresh milk smells like because we've smelt it thousands of times before. When we smell off milk we recognise it immediately. If the milk is only just turning sour and we are uncertain, we may decide, cautiously, to taste it to provide additional confirmation. This is exactly the process applied in detecting food fraud. One test may be enough, but sometimes a combination of tests are required to assess some characteristic, or combinations of characteristics, that can be compared to the known characteristics for that food. The known characteristics are obtained by testing many (usually thousands) of authentic examples of that foodstuff. This collection of authentic examples is known as a reference collection. The test results from the reference collection provide a database of the defining characteristics or standards (often within a range to account for natural variability) that are repeatable and reliable.

Let's consider tomatoes. At the mention of the name we picture them immediately as a red fruit with smooth skin, but the colour varies from green through orange to red depending on ripeness. Size also varies from cherry to beefsteak, and shape from spherical to plum. Once you know all this you can use your judgement (your own database) to decide whether the fruit on offer is a tomato. As to whether it's a vine-ripened tomato is totally bizarre; there seems to be a premium associated with having a piece of the vine still attached, even if that vine was no longer attached to the plant while the tomatoes ripened. It reminds us it came from a plant and smells quite nice, but we immediately throw the vine into the recycling bin. However, vine or otherwise, few of us will be aware of any criteria on which to judge whether a tomato is organic or locally grown. This information requires knowledge of characteristics that lie outside our regular frame of reference, and, more seriously, is beyond any capabilities we have to make such judgements. At this point we have to rely upon labelling and this is where opportunities open up for the fraudster. Questions concerning how food was produced and where it was produced are among the most challenging facing food testing laboratories.

The guiding principles of food fraud detection

The detection of rogue foodstuffs, or rogue ingredients within foodstuffs, relies upon the identification of a distinguishing physical or chemical/biochemical characteristic – a 'fingerprint' or 'signature' – that sets the adulterated foodstuff or ingredient apart from the accepted characteristic(s) of a given foodstuff or ingredient. Fingerprinting is a relatively simple idea used in the environmental, biomedical and forensic science field. It recognises that variation in chemical composition between different sources of a material can be used diagnostically to distinguish them. In the biomedical area, perturbations in metabolic biochemical 'fingerprints' of breath and urine are increasingly being promoted to spot different diseases, especially in early diagnosis. In the forensic field the different chemical compositions of illicit drugs, explosives and firearm residues can be used to link evidence from a criminal to a crime scene or crime object. Exactly the same idea is used in detecting food fraud, but there are some specific considerations in analysing food:

- The *chemical complexity* of foods. These are biological materials, animal and plant products, which are made up of highly complex mixtures of thousands of biochemically complex compounds, such as carbohydrates, proteins, lipids and nucleotides, in addition to water, minerals and trace elements. This complexity offers opportunities to both the fraudsters and the analysts. Adulterants can be hidden in complex mixtures; however, complex mixtures can also provide highly characteristic fingerprints.

- The *inherent variability* in the compositions of biological materials. Mother Nature's imperfections unwittingly aid the fraudsters. The compositions of even the same foodstuff are inherently variable. This natural variation is all too familiar to biologists, especially ecologists, biogeochemists and geochemists who study the natural world. The variability is generally not large, sometimes only a few percentage points across the global range, but that's enough to open the door to exploitation by the most skilled fraudsters. This variability, combined with the biochemical complexity of foods, offers the fraudsters numerous opportunities to hide their rogue ingredients.

- *Food processing,* including refining, mixing and cooking, can cause deviations in the food 'fingerprint' from that of the raw ingredient(s). Refining sugar, for example, involves separating components of the original product: raw cane sugar is separated into molasses, pure crystals and impurities. Refining and processing can thus remove the fingerprint capacity of the very compounds that many tests target, such as proteins and DNA. The complexity of detecting fraud generally increases with the increasing number of ingredients and the complexity of processing methods used to prepare the food in question. The fraudsters love processed foods as they are the perfect medium for hiding rogue ingredients. For the analyst, this means that a thorough understanding of how the food is processed is necessary in order to frame the right questions for detection.

Chemical fingerprinting can operate at two main levels, bulk and individual compounds, and both have their virtues. Bulk analyses tend to be quicker and cheaper, but analyses at the individual compound, or molecular species, level can give highly specific fingerprint information that increases the chances of detecting differences in different source materials. For example, bulk analysis can determine the percentage of nitrogen in a product, which might be a useful first step in assessing protein content (to test whether milk has been watered down, for example). But tests for individual compounds would be needed to determine whether all the nitrogen present is in the form of protein or to confirm the type of protein present.

One way of thinking about fingerprints is to consider the password you use to access your computer: the greater the number and range of symbols you include, the more secure your password becomes. The same principle applies to chemical fingerprints: the more chemical information you use, the more characteristic the chemical fingerprint becomes. Many food fraud questions concern the protein(s) in food. Proteins are made up of 20 common amino acids, so detecting these and determining their concentrations would confirm protein was present and how much there was, but crucially not the type of animal or plant protein present. The power of chemical fingerprinting in relation to proteins lies in the sequence, or order, in which the amino acids are arranged in protein chains, which we will discuss later.

The honey case study

To look at how the detection of food fraud works in practice, let's work through an example with honey. It's a beautifully sweet, rather distinctive-tasting natural golden syrup, produced by the ever industrious honey bee, *Apis mellifera*. Honey bees collect the nectar of plants and convert it into honey as a source of energy for the colony. But careful tending by beekeepers ensures the bees produce a surplus for our delectation. Annual global honey production amounts to an astonishing 1.2 million tonnes. That's a lot of jars of honey, but seemingly insufficient to satisfy our needs. It's a sobering thought that one worker bee typically makes only 0.5g (0.018oz) of honey in her lifetime.

Premium or specialist honeys are expensive as they claim a range of beneficial nutritional and curative properties, which increase their popularity. However, specialist or not, any honey we buy should be pure honey. Demand and premium prices motivates fraudsters to stretch this precious hive product by adding cheaper ingredients, such as corn syrup, fructose, glucose, beet sugar and cane sugar. Indeed, honey is recognised by the European Parliament as among the top 10 foods most likely to be adulterated.

Surely testing the purity of honey will be simple enough? As you will see, nothing could be further from the truth. There are a number of descriptors that must be included on honey labels in order to accurately describe the product and all of these claims have to be true for the product to be authentic. The Codex Alimentarius Commission, established in 1963 by the Food and Agriculture Organization of the United Nations (FAO) and the World Health Organization (WHO), sets international food standards, guidelines and codes of practice. Its 12-page standard for honey describes nine different categories relating to differences in source or means of processing honey: blossom or nectar honey, honeydew honey, comb honey, chunk or cut-comb honey, drained honey, pressed honey, extracted honey, filtered honey and baker's honey. The country of origin and predominant plant the bees used must also be stated, as well as whether the honey was blended. What unifies all these commodities is that they are honey and any honey must be 100 per cent pure honey.

Compared to many of the foods we eat, honey is a relatively simple substance biochemically. It is a super-saturated sugar solution. Supersaturation means more sugar is dissolved in the water honey contains than is theoretically possible; that's why honey crystallises so readily in the jar. This crystallisation is quite normal as honey ages and is *not* an indication of adulteration. The processing of the nectar by worker bees involves repeated regurgitation and enzymatic conversion to honey. The familiar viscosity of runny honey develops through the regurgitation process with the worker bees evaporating just enough water using their wings to make the consistency of the honey deposited in the cells of the honeycombs just right.

The resulting honey is composed of two main sugars, fructose and glucose, with a smaller proportion of sucrose and maltose. Together these sugars account for around 80 per cent of the total mass of a jar of honey. The remainder is largely water, though this is never greater than 20 per cent, otherwise the honey will ferment. There are also tiny amounts of vitamins and other organic compounds, including traces of proteins and minerals. The chemical composition of honey varies depending on the nectar collected by the bees, and it will change throughout the season and will vary by region depending on what flowers are in bloom. Often the aromatic compounds in the nectar are carried over to the honey; it's quite easy to tell when honey has been made by bees foraging in lavender, for example, because the honey has hints of a lavender aroma. Beekeepers have a good sense of where their bees are foraging given what's in bloom within a certain range of each hive, but it's not an exact science as bees are free-flying insects. As we will see, many of these properties can be used as criteria in honey authentication.

Years of research have sought to establish the chemical composition of honey in order to protect the producers. Compilation of the known compositions of thousands of samples of honey has allowed the global ranges for the physical characteristics and chemical components of honey to be defined. The Codex Alimentarius standard lists detailed information on the following criteria for specified honeys: sugar content (including limits for fructose, glucose and sucrose), moisture content, water-insoluble matter content, electrical conductivity, diastase enzyme activity and hydroxymethylfurfural (an organic compound) content. Without going into the details of all of these, most of them contribute to the chemical fingerprint information of honey – they are the complex secure password that confirms the purity of the product.

Fingerprinting: how pure is our honey?

So our first forensic question is whether our jar of honey is pure. The major criterion used to authenticate the purity of honey is the sugar composition – the honey sugar fingerprint. Sugars not only make up the bulk of the honey, but the relative proportions of individual sugars are fairly consistent worldwide. The technique central to measuring mixtures of compounds, such as sugars, is chromatography. You may have had an opportunity at school to separate the different colours in ink on a filter paper. If not, here's a quick refresher. Filter paper that the students have marked with a black marker is placed upright in a glass containing just enough water to wet the bottom edge of the paper. As the water travels up the filter paper, the pigments left by the black marker are carried along for the ride. The different coloured pigments will travel at different rates and will therefore separate out, turning a black mark into a rainbow of colour. That's paper chromatography. And like all chromatography, it's based on the principle that the mixture to be analysed is carried in a mobile phase (either a liquid or a gas) through a stationary phase (usually solid). The different constituents in the mixture travel at different speeds and will be retained in different spots on the stationary phase. Where the components separate out on the stationary phase can then be compared with known substances to help resolve their identities. Honey sugars can be determined by gas chromatography (GC) but it is more easily done using high-performance liquid chromatography (HPLC). This technique forces mixtures of compounds dissolved in the liquid mobile phase through the stationary phase at high pressure, which makes it faster and gives it superior resolving power for a wider range of compound classes, making it one of the most widely used separation techniques in analytical chemistry.

As shown in Figure 2.2, HPLC separates the sugars in honey into fructose, glucose and sucrose, but also quantifies the relative proportions of each – this defines the honey sugar fingerprint. Pure honey will consistently contain these sugars within the following ranges:

- Fructose: 31.2 to 42.4 per cent;
- Glucose: 23.0 to 32.0 per cent;
- Sucrose: 0 to 2.8 per cent.

Honeys with sugar compositions outside these ranges are therefore highly likely to have been adulterated. Honey's

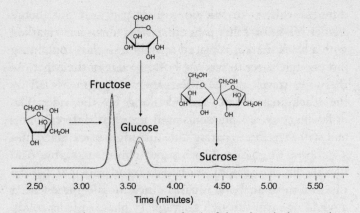

Figure 2.2. Chromatograms (overlain) of shop-bought honeys show how consistent the composition is of the three main sugars (structures shown). The areas under the peaks are proportional to the amount of the sugar present.[1]

low abundance of sucrose (the familiar cane or beet sugar is sucrose) makes this a poor choice of adulterant as adding even a small proportion would be easily detected – yet it still happens. However, the ranges for fructose and glucose are substantial, around 10 per cent, and their overall abundances are both high. Therefore, despite HPLC providing a valuable chemical fingerprint, there is plenty of scope for adding cheaper sugars or mixtures of sugars, which might never be picked up by this approach.

A very troublesome adulterant in honey is high fructose corn syrup (HFCS). This is produced in the US as a replacement for sugar by treating corn (maize) starch with enzymes that convert some of the glucose to fructose. A range of HFCSs are produced, with one, HFCS 55, having a remarkably similar fructose and glucose content to honey. So even if the results of the HPLC are consistent with a pure honey, this does not necessarily mean the forensic analysis is over, particularly if there is evidence, such as questionable paperwork, which caused suspicion of the honey to begin with.

Stable isotopes to the rescue!

So where do we go from here with authenticating our honey? The sugar fingerprint is consistent with a pure honey, but there's cause for suspicion nonetheless. The next question may be to determine whether HFCS, or indeed any corn syrup, has been used. This time nature is on our side as we look to answer this question. There are two major groups of plants in the world called C_3 and C_4 plants. The C_3 and C_4 terms refer to the different ways in which the plants capture carbon dioxide (CO_2) from the atmosphere during photosynthesis in order to make their sugars. The C_4 pathway was an adaptation among some plant groups to high light intensity, high temperatures and dry conditions as the pathway is more water-efficient. CO_2 exists in two forms that are important here: $^{12}CO_2$ and $^{13}CO_2$. They are very similar, differing only in the fact that the carbon atom (C) in $^{13}CO_2$ contains an extra neutron (^{12}C and ^{13}C are the stable isotopes of carbon). The extra neutron in ^{13}C does not make a big difference and $^{12}CO_2$ and $^{13}CO_2$ follow each other around, behaving in the same way. However, the extra neutron makes $^{13}CO_2$ physically heavier than $^{12}CO_2$. Most of the time this is of no consequence; however, the enzymes involved in capturing CO_2 in photosynthesis in all plants are sensitive to this small difference in mass and use $^{12}CO_2$ over $^{13}CO_2$. The result is that the sugars made by plants through photosynthesis have proportionally less ^{13}C than the atmosphere. Furthermore (and this is absolutely critical), C_3 plants contain less ^{13}C compared to C_4 plants due to a difference in the way they capture CO_2. And this can be used to beat the food cheats.

Fortunately, the vast bulk of the honey produced in the world is produced in regions where C_3 plants predominate within the ecosystems. Thus, based on the 'you are what you eat' (or in this case, what bees chew and spit out) principle, the honey the bees produce will have a C_3 plant fingerprint. In contrast, the corn (maize) plant is a C_4 plant, so any added HFCS will have a C_4 plant signature. So if you start mixing HFCS into honey, the ratio of $^{13}C/^{12}C$ increases as more corn syrup is added. This can be readily determined using an isotope ratio mass spectrometer (IRMS), which is a very sensitive weighing machine capable of determining the precise ratio of $^{13}C/^{12}C$ in biological materials, including foodstuffs. There is, however, one catch: beekeepers often feed their bees on pure sugar, especially during colder periods when they don't fly. Sugar cane is also a C_4 plant, so any legitimate honey produced when it's being fed to the bees will have a $^{13}C/^{12}C$ ratio similar to HFCS. This would suggest adulteration, when the beekeeper is simply using a perfectly acceptable practice to maintain his or her hives. Here's a perfect example of why food analysts need to have a thorough understanding of how the food is made.

Realising this problem, a modification of the stable isotope test was proposed. The fraudulent addition of HFCS can be detected by comparing the $^{13}C/^{12}C$ ratio of both the honey sugars and the honey protein. The $^{13}C/^{12}C$ ratio of the sugars and the proteins in a genuine honey are very similar, and the small difference is relatively constant across all honeys. When bees are fed the C_4 cane sugar, this is incorporated into both the sugars and the proteins of the honey as the bees process it. However, if HFCS is added to the honey after bee processing, the carbon isotope composition of the sugars changes but that of the protein doesn't. Sufficient differences in the isotope values unambiguously confirm adulteration and can be quantified down to a 7 per cent level of added HFCS.

By judicious use of two types of analysis we could be fairly confident that the honey we are about to spread on our toast is indeed pure honey, but what about other claims on the label?

The mānuka mystery

Perhaps our hypothetical jar is labelled as mānuka honey, which is extremely expensive. One UK supermarket currently sells raw mānuka honey for £39.95 (US$63) for a 340g (12oz) jar. Mānuka honey is produced by bees foraging specifically on the mānuka or tea tree (*Leptospermum scoparium*), which grows uncultivated throughout New Zealand and southeastern Australia. Research suggests that it has a number of curative properties, particularly outstanding antibacterial activity. As a result of the premium price, the adulteration of mānuka honey has reached astonishing proportions. It is reported that the majority of mānuka-labelled honey on supermarket and health food shop shelves worldwide is adulterated.

The New Zealand Unique Mānuka Factor Honey Association (UMFHA) estimates that just over 1,500 tonnes of pure mānuka honey are made annually in New Zealand. It is truly remarkable, therefore, that an estimated 9,070 tonnes of honey labelled as mānuka is being sold around the world annually; more than five times the amount of mānuka honey produced is sold. The UK alone consumes around 1,800 tonnes each year. Something's dreadfully wrong somewhere!

Analyses performed by the UK FSA between 2011 and 2013 showed the majority of mānuka honeys sampled lacked the desirable non-peroxide antimicrobial activity deemed to be exclusive to true mānuka honey. While it is recognised that the antimicrobial activity can diminish between production and sale, the disparities between the true production and sales tonnages is truly mind-boggling. In response to this, in 2014 New Zealand news reported that the Ministry for Primary Industries had begun working on a new guideline for mānuka honey. To support this guideline, the UMFHA was reported to be collecting honey samples from around the country to establish a chemical profile of mānuka for use as a fingerprint and reference point in honey testing.

So how can the tests outlined above be taken a step further to confirm unambiguously that our honey is truly mānuka as it claims on the jar? This is an exceptionally challenging problem that the food forensic scientists continue to explore. The traditional approach is to test the pollen content to confirm the forage species. The standard for mānuka honey is only 70 per cent *Leptospermum* pollen, but pollen analysis is complicated by the fact that kānuka (*Kunzea ericoides*) is a tree from the same family found in the same habitats, which has identical pollen to mānuka. Likewise, the Australian native jelly bush (*Leptospermum polygalifolium*) is classified to the same genus as the mānuka. The bottom line is that a new test is needed to separate these different honeys if the unique classification of mānuka honey is to be upheld.

Omics and the food testing revolution

To get to the floral origins of mānuka, as well as other aspects of honey authentication, we turn to omics-type approaches. 'Omics' is a scientific neologism (expression) given to the fields of study in biology ending in *-omics*, namely genomics, proteomics and metabolomics. Given that the vast majority of food derives from living things (we hope), the use of omics approaches is heralding a new dawn in twenty-first-century food fraud detection in many areas. Overall, omics research aims at *the collective characterisation and quantification of biological molecules that translate into the structure, functions and dynamics of living organisms*. The goals of food fraud detection might not appear to fit comfortably within this definition. However, the most basic question addressed in the authentication of a food is species origin, and it is here that the principles of omics and the demands of food fraud detection converge.

First, a few words on the basic principles of omics. All organisms on Earth are the product of their DNA sequences. The building block chemicals for DNA are called nucleotides and the sequence in which they appear is called the genetic code. Genomics is the study of the structure, function and evolution of these sequences. Sections of the genetic code, known as genes, are the blueprint for building proteins. Proteins are made up of chains of amino acids. Proteins differ from one another primarily in their sequence of amino acids, their sizes and shapes. While we commonly associate proteins with the structural and mechanical material that makes up muscle, they are also responsible for catalysing metabolic reactions (enzymes), replicating DNA and transporting molecules from one location to another (for example, blood haemoglobin is the protein that carries oxygen). Proteins are also sensitive detectors in responding to stimuli from inside and outside the body (for example, a hormone receptor on the cell surface is a protein). The study of the structure and function of proteins is proteomics. The final omics technique in the toolkit for food fraud detection is metabolomics. While genomics and proteomics approaches are based on the DNA and proteins, metabolomics essentially looks at all the other processes happening in the cell. An example might be in toxicity screening. Urine, which contains the waste products from cellular metabolism, can be analysed to see what physiological changes a toxin has caused in the body. This is a different approach from looking at organs individually to see how the toxin has affected them. We'll return to metabolomics and genomics shortly, but to verify our mānuka honey we're first going to see if we can use proteomics.

The potential of proteomics

We have mentioned that the stable carbon isotopic composition of the protein in honey can be used to detect HFCS added to honey, but there is much more to proteins than their stable isotope compositions. Indeed, an emerging technique in the food fraud detection toolkit is proteomics. The amino acid sequences of proteins can be species-specific and thus can be used to define the species origins of unknown proteins. So why use proteins and not DNA to assign the origin of a foodstuff? First, the equipment necessary to analyse proteins is more common, whereas DNA sequencers are not necessarily standard equipment for many analytical labs (though this is changing as these methods become cheaper). Second, proteomics can have some major advantages when verifying the origins of highly refined or processed animal or plant products.

The proteomics approach has been around for more than 20 years in biochemistry. The method involves treating an unknown protein or mixture of proteins with an enzyme called trypsin, which cuts the proteins to produce a mixture of smaller fragments called peptides. Trypsin recognises particular amino acid sequences and clips the protein in this region every time, so it is predictable and reproducible. The resulting mixtures of protein fragments are separated by HPLC and analysed by mass spectrometry (MS), to determine the amino acid sequences of the peptides. The MS effectively weighs the individual peptides, splits them further and measures how much of each peptide fragment there is and the order of the amino acids they're made up of. Plotting the weights (molecular masses) and how much (abundances) of each peptide gives highly diagnostic fingerprints. These can then be compared with databases of peptide sequences from known proteins from candidate food species; assignments of species of origin can be made with a very high degree of confidence. We will see later how this approach can be especially effective in detecting meat fraud, but can it be used to fingerprint species-specific proteins in our honey?

Pier Giorgio Righetti and his team from the Politecnico di Milano, Italy tried to isolate the proteins in honey from chestnut, acacia, sunflower, eucalyptus and orange.[2] Sadly the protein content of the honeys they tested were many times lower than had been previously reported. Even more disappointing, their proteomics analyses showed that all but one of the proteins they could identify in the honey were derived from the bees themselves – these were enzymes in the bees' saliva. Their attempts at identifying plant (pollen, nectar) proteins were unsuccessful, so at this stage honey proteomics seems to be a non-starter for floral source identification. We'll have to turn to other approaches to verify the origins of our mānuka honey.

Mesmerising metabolomics

As plants and animals go about their daily functions associated with living – breathing, eating, excreting, to name a few – they produce different mixtures of compounds. Differences arise in these compounds depending on the organism and where in the world it's living. We should, in theory, then be able to use sophisticated analytical techniques to produce a high-resolution chemical fingerprint of these compounds to determine whether our honey was produced from mānuka trees in their endemic regions of New Zealand and Australia.

One of the techniques used in metabolomics is nuclear magnetic resonance spectroscopy (NMR), which many will be familiar with in relation to medical diagnosis. The method actually has its origins in analytical chemistry laboratories where it is mainly used to investigate the structures of organic molecules. It works through the interactions of radio frequency radiation with the atoms of molecules in a magnetic field. In 2012, Italian-based researchers showed how 600MHz ^{1}H NMR (600MHz refers to the resonance frequency used and ^{1}H refers to the nuclei being studied) could be used to discriminate honeys of different botanical origins.[3] Over a two-year period they collected NMR spectra of 353 extracts of monofloral Italian honey (such as acacia, linden, orange, eucalyptus, chestnut and honeydew), as well as polyfloral honeys. They identified specific markers for each monofloral honey, then used NMR-based metabolomic fingerprinting combined with multivariate statistics to discriminate between the different types of honey. The method required very little sample preparation; it was fast, reproducible and is claimed to be more objective than melissopalynological (pollen) analysis. A year later, a different group of researchers identified 13 metabolites in honey that had at least one unambiguous resonance using ^{1}H NMR that could be used to quantify the compounds, including carbohydrates, aldehydes, aliphatic and aromatic organic acids.[4] They used the technique to quantify these compounds for mānuka honey, but didn't attempt to use it to discriminate mānuka from other types of honey.

Another metabolomics approach is to target the volatile compounds (organic chemicals with a high vapour pressure at room temperature) present in the aroma of honey as a means of verifying their geographical origin. The volatiles are collected from the air above the honey using a technique called solid phase micro-extraction (SPME). SPME uses a polymer-coated fibre that absorbs the honey's volatile components. The volatiles can then be separated using different chromatography techniques and analysed using different methods of mass spectrometry. These data from the volatile components of the different honeys are used to construct a model that can classify the geographical and floral origins of those different honeys. The model can then be used to take unknown honeys and predict their origins.

In 2014, a research group from Dresden, Germany applied SPME to analyse the volatile compounds in mānuka honey as well as the pollen-identical kānuka honey and closely related jelly bush honey.[5] They combined their findings for the volatile compounds with an analysis of the non-volatile compounds using HPLC and mass spectrometry. They investigated the complex chemical signatures for the eight samples of mānuka honey, seven samples of kānuka honey and one sample of jelly bush honey using chemometrics approaches; *i.e.* by using advanced statistical methods they were able to pinpoint characteristic substances for each honey, which allowed classification of each honey into one of three clusters. Though the model enabled them to correctly classify each of the honeys they tested, models are only as good as the data input. The researchers admit that although their approach has potential to be used for routine honey authentication, their sample size was small. A larger database of honeys collected from different production years and from different regions in New Zealand is required before these methods can be used to authenticate our mānuka honey, but metabolomics seems to hold great potential in honey authentication.

DNA: the ultimate fingerprinting approach?

Perhaps genomics can resolve our mānuka mystery. The unique genetic code for each species can be exploited in food fraud detection as a genetic fingerprint rather than a chemical fingerprint. The major difference between a simple sugar composition and a DNA sequence lies in the amount of data available in the latter. If we go back to the password analogy, DNA gives the ultimate secure password due to its specificity.

Labs all over the world sequence areas of DNA of interest and entire genomes for different species all the time. And while this is helpful in answering specific questions, the lack of standardisation isn't particularly useful in terms of building a repository of data for species around the world; in other words, a database that could be used to identify all species, including those we eat. In 2003, Paul Hebert's research group at the University of Guelph, Canada, proposed DNA barcoding as a method of quickly identifying all living things on Earth. Barcoding would target a standardised region of the DNA. Just as barcodes, such as the Universal Product Code (UPC) or International Article Number (EAN), are used to identify retail products, a short sequence of DNA can be used to identify species – except, instead of a series of numbers it's a series of nucleic acids. This need to identify species is as fundamental to biological studies as the Periodic Table of Elements is to chemistry. However, unlike elements, species are going extinct faster than they can be identified ... and there are millions of them! Hebert and others proposed barcoding as a fast and simple method that could meet this urgent need to differentiate living organisms.

A UPC symbol, in its most common format, has 12 digits with 10 possible values at each digit (zero to nine). This provides one trillion (10^{12}) possible combinations. DNA has four possible values: the nucleotides adenine (A), guanine (G), thymine (T) and cytosine (C). A gene 500 nucleotides long could theoretically provide 4^{500} unique codes – more than enough to assign all living species a unique code.

But which segment of DNA makes the best barcode? The segment needs to be as universal as possible among different taxonomic groups and reasonably short for efficiency reasons. It needs to be easily recognised, so it has to have a region of DNA on either side of it in the sequence that doesn't vary much between species. This conserved region acts like a bookmark when the sample is processed. The segment itself, however, needs to be variable enough that it can be used to differentiate species, but not so variable that it differs within a species. Much like Goldilocks's porridge, the mutation rate can't be too fast or too slow ... it has to be just right. In plants, which is what's relevant to our honey example, the segments of DNA that are targeted for barcoding are two gene regions in the chloroplast, known as *matK* and *rbcL*.

The simple version of the process is that DNA is extracted from the sample and then put through a process to amplify the target gene segment – known as Polymerase Chain Reaction (PCR). The sample is then run through a DNA sequencer and what's returned is a string of As, Gs, Cs and Ts – the barcode. These data can then be uploaded to the Barcode of Life Database (BOLD) to be compared with existing barcodes that have been collected from expert-verified specimens. The unknown specimen will either match an existing sequence or come back as being divergent enough from anything else in the database that it is considered new to the database. The methodology is streamlined and relatively inexpensive and a single DNA barcoding facility can process hundreds of thousands of samples each year at a cost of about £6.31 (US$10) per sample, not including labour and supplies.

The reference database is a critical component to the Barcode of Life project. As well as providing sequence information (the barcode) and species name, there is additional information available regarding the quality and source of the sample and the reliability of the identification. A specimen from a botanic garden that has been identified by the curator, for example, is likely to be more reliable than ancient DNA harvested from a partial grain extracted from a sediment core pulled out of the bottom of a lake. This secondary information helps those using the data to establish a level of confidence in the results. While this is important in communicating scientific findings, it also helps to establish the quality of the evidence if used to help prosecute in a case of food fraud.

So can we use DNA barcoding to authenticate the plant origins of our honey? Back in 2010 Tom Gilbert and colleagues at the Natural History Museum, University of Copenhagen, Denmark, showed that even small samples of honey, one millilitre, contain enough DNA for effective PCR analysis.[6] They tested a number of honey samples and recovered sufficient lengths of DNA to allow them to identify taxonomic groupings of insects and plants. Interestingly, DNA from insect mitochondria (the organelles responsible for energy production in the cell) dominated over plant organelle DNA (DNA from parts other than the nucleus, such as the chloroplast). Plant nuclear DNA was present at the lowest levels. As the target areas for barcoding are in the chloroplast DNA, this confirmed that DNA could be used for characterising the floral source of honey.

At the time of writing this chapter, Massimo Labra and his group at the Università degli Studi di Milano-Bicocca, Italy, have recently put barcoding honey to the test.[7] They analysed four multifloral honeys produced at different sites in the northern Italian Alps, using the *rbcL* region as well as one other region as barcode markers. They assembled a DNA barcoding database for the region where the bees were foraging – a total of 315 species. From the DNA they extracted from their honey samples, they were able to identify 39 plant species in the four honeys, including *Castanea* (chestnut), *Quercus* (oak) and *Fagus* (beech) trees, as well as many herbaceous taxa. An unexpected finding was the DNA from a toxic plant, *Atropa belladonna*. Though it's unlikely that the honey contained any toxic metabolites of the plant, it shows the potential usefulness of DNA barcoding for evaluating honey ecotoxicity and other food safety-related issues.

As we've already stated, the strength of barcoding lies in the database. There are currently seven barcoded specimens of the mānuka tree (*L. scoparium*) in BOLD, which have been mined from the GenBank database. There is very little secondary information associated with the information in the database. There is one sample that was collected by the Royal Botanic Gardens, Kew (UK), so this is a likely candidate as an expert-identified specimen. In theory, we could therefore send a sample of our honey off to a lab for analysis to see whether it contained only plant DNA from the mānuka tree.

Where does my honey come from?

If we can use barcoding to verify that our honey is from the mānuka tree, then it's rather obvious where the honey has originated as the tree is only found in a restricted area of the world. However, what if we want to verify the geographic origins of a honey that's either polyfloral or made predominantly from a flower that's more widespread geographically? Knowing the origin of honey is important in understanding the risk of adulteration as well as contamination. Some countries have a history of adulterating honey. And less stringent pesticide and antibiotic regulations in some countries make their honey more prone to contamination with the residues from these chemicals. Being devious about the geographical origin of honey can present a health risk.

The movement of food around the world, often thousands of miles away from its site of production, means we are moving food to regions characterised by different ecosystems, climates and soil types. Remember the 'you are what you eat' principle? Plants and animals growing in different regions naturally record in their tissues the chemical and stable isotope signatures of their environment. As plants grow, they take up nutrients from the soil and atmosphere. Where they grow will be manifested in differences in the macro- and micro-nutrients they absorb due to different soil types (reflecting underlying geologies), fertilisers, rain and soil water compositions. The chemical and isotopic compositions of all plants will be directly affected by these differences in composition. The chemical and isotopic signatures recorded in the plants are then passed on to the consumers, notably grazing animals (including bees), when they eat the plants and assimilate plant nutrients into their tissues. Thus, all the food we eat, plants or animals, possesses a chemical and isotopic signature of where it was produced. For the most part, we can't detect these chemical and isotopic signatures based on the taste or texture of the foods and beverages we consume – though a honed palate is arguably able to assign wines to certain growing regions. Decades of ecological and biogeochemical research have provided us with a detailed understanding of the major factors that affect the chemical and isotopic compositions of plants and animals.

Recognising these geographic signatures, the EU funded a major initiative known as the food traceability project (TRACE). It brought together researchers from Germany,

Italy, the UK and Austria, to investigate if it was possible to discriminate between honeys produced in different parts of Europe with different climates and geological conditions. More than 500 authentic honeys were collected from 20 different regions. The honey protein was isolated and the stable isotopic compositions were determined for carbon ($^{13}C/^{12}C$), hydrogen ($^{2}H/^{1}H$), nitrogen ($^{15}N/^{14}N$) and sulfur ($^{34}S/^{32}S$). These are the major elements that make up protein. The lighter isotopes (the ones in the pair with the smaller superscripted number) are more abundant naturally. And just as ^{13}C is the heavier version of ^{12}C with its extra neutron (recall the differential uptake of $^{13}CO_2$ and $^{12}CO_2$ by C_3 and C_4 plants), ^{2}H, ^{15}N and ^{34}S are heavier versions of these elements. These stable isotopes can record different invisible processes going on in the environment, just as they did in the different photosynthetic pathways of plants. In fact, this gives isotope biogeochemists a different view of the world. For them, rain doesn't just bring vitality to the garden or thwart picnic plans. For the isotope biogeochemist, rain is painting the environment with different isotope compositions – an isotope landscape if you will. Water molecules that contain the lighter hydrogen isotope ($^{1}H_2O$) evaporate more readily from the surface of the sea than the heavier water molecules ($^{1}H^{2}HO$). As vapour in the air condenses into clouds and moves from the sea over land, $^{1}H^{2}HO$ will precipitate out first and the lighter water molecules will be carried further inland. As a result, the hydrogen isotope composition of the clouds changes in a predictable way and there is geographical fractionation of the isotopes across the landscape. Rain falling increasingly far away from the coast contains increasing proportions of $^{1}H_2O$ relative to $^{1}H^{2}HO$. Plants (and animals) drink this water and the hydrogen isotope ratio for their region is incorporated into their tissues.

Together, isotope biogeochemists and food fraud detectives can exploit the stable isotope fingerprints recorded in plants to devise new ways of assigning the geographical origins of different foods. In fact the predictability of stable isotope fractionation across the landscape – known as an *isoscape* – has resulted in the generation of maps showing how the stable isotope compositions of elements vary across different regions of the world due to variations in rainfall, temperature, geology, land use, plant types and other underlying processes. These maps can help to verify claims of the region of origin displayed on food labels. Of course, a fraudster could in theory fake a food product that would make it isotopically identical to the real thing, but achieving this would be so costly that the economic gains would make it nigh on pointless.

Results from the EU-funded TRACE project showed that the stable isotope compositions – carbon, nitrogen, hydrogen and sulphur – in honey protein could be used to verify the origin of honey.[8] Carbon and sulphur isotopes were particularly useful in confirming the origins of the European honeys tested. Other research has indicated that the utility of isotopic analysis varied depending on the honey component under investigation. The hydrogen and oxygen isotopes measured in pollen, for example, don't seem to reflect regional precipitation isotope compositions as well as these isotopes measured in liquid honey or beeswax. There are also certain limitations with this approach because if the honey is shipped out of the region for processing, then moisture in the local atmosphere will exchange with the water in the honey, changing the isotopic signature. If the honey is unfiltered though, the hydrogen and oxygen isotopes could be measured in any beeswax retained in the honey. Beeswax, unlike liquid honey, retains the hydrogen and oxygen isotope fingerprints of where the honey originated. Of course, the intelligent fraudster will know this, and may potentially add beeswax to trick the test.

There are victories, but the game keeps changing
Food fraud detection is evolving in an ever-changing landscape. There is an increasing number of commercial foodstuffs on the market, each with inherent biochemical

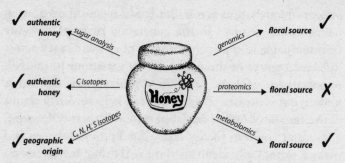

Figure 2.3. Maybe I can trust my honey if it's tested using the very best analytical methods science has to offer!

complexity. The scale and complexity of food supply chains is unprecedented. And for every new method of food authentication described in the literature, there are at least 20 new methods described for processing, preserving and manipulating food. The food cheats know this, and design their fraudulent methods within this world of complexity, uncertainty and chaotic practices.

As our honey case study shows, there are some very advanced analytical chemical and biochemical methods that, when combined with informatics and chemometric techniques, are pointing the way forward. These new methods make for a well-equipped toolkit. Recent food fraud scandals have provided the impetus for governments around the world to direct financial and human resources into putting these toolkits to use. There has even been enough momentum to motivate concerted cooperative cross-border monitoring and control initiatives. Collectively, this means that our capacity to make serious dents in the activities of food fraudsters is now more advanced than it ever has been. Yet, why does it feel as if we are still playing catch-up? As the methods we've described in this chapter mature and we become more confident that our supermarket honey is what the label states, what new ways will fraudsters find to make their money? This is of particular concern at a time when pollinators (not just honey bees) are in crisis around the world.

In 2001, the US Department of Commerce implemented anti-dumping duties (a 300 per cent tariff) on honey imported from China to discourage suppliers from flooding the US market with cheap adulterated honey and putting their own domestic beekeepers out of business. To get around this duty, some US importers simply purchased honey from intermediary countries – 'honey laundering', so to speak. All honey labelled from China is subjected to veterinary testing in the US due to the likelihood of pesticide and antibiotic (some of which are banned) residues in the honey. If it's laundered through another country and does not state that it's a product of China, it will not be subjected to these tests. To get around this, all honey could be tested, but this is simply impractical. Honey from China also tends to be ultra-filtered so there is no pollen remaining in the honey at all. Pollen carries many of the benefits of honey, and the US FDA has ruled that honey that contains no pollen is no longer honey. It also makes origin testing more difficult. In 2011, *Food Safety News* purchased 60 jars of honey from around the US and had them tested by a pollen expert at Texas A&M University. Seventy-six per cent of the samples from supermarkets and cash-and-carry stores like Costco contained no pollen whatsoever. Is this even honey at all?

The solutions to this are both complex and simple. The complex solution for globalised suppliers is to commit to selling authentic products by signing up independent third-party auditors committed to socially responsible corporate practices. In the US, the True Source Honey organisation has created a certification programme that allows parties involved in the supply chain to demonstrate compliance with food purity and safety regulations. That's the complex solution. The simple solution is to avoid the foreign imports by buying honey from your local beekeeper.

Both of these solutions will probably increase the price we have to pay for honey. It's a sad day when you have to pay a premium to know that what you're eating is what you want to eat!

CHAPTER THREE

A Slippery Deal

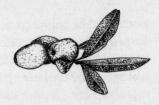

One of the biggest food frauds ever perpetrated involved vegetable oil. Although it was actually atypical of the usual adulteration or mislabelling of vegetable oil, the sheer scale is worthy of recounting. The incident became known as the Salad Oil Scandal or Soybean Scandal, and was exposed in 1963, causing over US$150 million (£96.5 million; more than US$1 billion, £643 million, at present-day rates) of losses to such corporations as American Express, Bank of America and Bank Leumi, in addition to many international trading companies. The scandal has been likened to the 2007–8 subprime mortgage crisis. The financial ramifications illustrate the importance of vegetable oil as a global commodity.

The Salad Oil Scandal was an astonishing fraud perpetrated by the Allied Crude Vegetable Oil Refining Co. in New Jersey, led by Tino De Angelis. The idea was very simple: load ships with water and float a few feet of soybean oil on the surface – oil floats on water. Inspectors looking into the tankers thought they were full of oil and certified the shipment. De Angelis then used the certificates to obtain massive loans from Wall Street banks and companies, based on the bogus amounts of soybean oil confirmed by the inspectors to be in the ships. The quantities incidentally suggested the existence of much more salad oil than was actually accounted for in the entire US at the time. The scam was exposed in November 1963 and resulted in chaos in the futures markets with the entire value of loans being wiped off the markets in minutes. The fraud was overshadowed by the assassination of US President John F. Kennedy on 22 November 1963. However, De Angelis eventually ended up serving a seven-year jail sentence. Although no science was used to reveal this fraud, the principle of production/supply versus sales auditing – as we saw in the previous chapter with mānuka honey and we will see again in later chapters – can be a critical step in exposing major food fraud.

Where there's big money, there are big cheats!

The massive scale of global vegetable oil production is what makes the vegetable oil trade such a lucrative business, moving vegetable oil fraud well up the food cheats' agenda. We're talking big numbers here – current global annual production for all vegetable oils stands at around 170 million tonnes and is growing year-on-year. The big four are palm, soybean, canola (also known as rapeseed) and sunflower oil, in that order. At least 50 other vegetable oils are produced for human consumption from different seeds and nuts, originating from various parts of the world, mostly the tropics. Seed oils make up the major part of edible oil production, while nut oils tend to be popular owing to their characteristic flavours and are used as specialist gourmet ingredients. The vast proportion of vegetable oil produced is for human consumption. Only a small proportion is used to generate biodiesel and much of this is recycled from spent cooking oil. The scale of vegetable oil production is a response to consumer demand, which can be surprisingly high and is set to rise as Mediterranean-style diets increase in popularity. In 2014, the annual per capita consumption of vegetable oils in Italy reached a high of 28kg (just shy of 62lb); that's the best part of 100ml (3.5fl oz) per person per day.

Getting clear on oil

The physical property that unifies all vegetable oils is that they are ... oily. When you buy a bottle of vegetable oil you're buying a bottle of fat – there are more similarities between vegetable oils and animal fats than you might

realise. Generally, oils are liquid at room temperature while
fats are solid. The liquid oils used for frying and preparation
of salad dressings are oils, while lard, butter and hard
vegetable fats, such as cocoa and coconut, are fats.

But oils can be made up of different materials: mineral
oils are derived mainly from petroleum; others, such as
silicone, are made from synthetic polymers; and the oils we
eat come from plants and animals. All of these chemically
different materials could be mixed together and you would
still have a clear oil. And indeed there are examples where
such substances have been mixed together in vegetable oil
adulteration. It's pretty revolting to imagine dressing your
salad or frying your food with a mixture of vegetable oil
and substances we are more familiar with as components of
car engine oil or floor polish. Because such substances
occur naturally and are permitted additives or trace
contaminants from production, limits are set by the
regulatory agencies for how much of these foreign
substances are allowable in vegetable oils. However, some
of these substances are intentional additions to change the
properties of the oil.

Reading oil product labels can be quite illuminating,
especially since new labelling regulations introduced in 2014
demand explicit listing of the plants used and other
ingredients added. For example, on reading the ingredient
list on a bottle of vegetable oil you may find dimethylpolysiloxane listed as an added 'antifoaming' agent.
Defoamers and antifoaming agents are added to liquid foods
to reduce the surface tension and therefore inhibit the
formation of foam. Dimethylpolysiloxane is an industrial
chemical probably best known as the silicone oil used in
breast implants. Amazingly, a US patent was lodged in 1991
(US4983413A) proposing the use of this class of industrial
chemicals to produce low-calorie fried foods and salad
dressings. Luckily it lapsed for failure to pay maintenance
fees. Since our digestive system cannot absorb or digest
these silicone compounds in the same way as normal

vegetable oils and animal fats, the patent proposed that portions of the oil content in low-calorie foods be replaced with organopolysiloxane compounds, achieving that fatty flavour without all the extra calories. Interestingly, McDonald's confirm on their website that 'The oil we use for our fried menu items, like our Chicken McNuggets, World Famous Fries and Crispy Chicken sandwiches, does contain a small amount of dimethylpolysiloxane. This is an FDA-approved ingredient that helps prevent the splattering of oil as foods are cooked.'[1] Dimethylpolysiloxane is an inert chemical, which makes it the best choice for breast implants, but it doesn't necessarily make it a palatable addition to edible oils – we'll take a little foaming and splattering instead any day!

The colour variations in pure vegetable oils arise from traces of plant pigments, such as polyphenols, carotenoids and chlorophyll, extracted from the seeds and nuts during refining. Despite some variation in colour, vegetable oils are all very similar chemically. All vegetable oils (and fats) are made up of triacylglycerols (TAGs), also known as triglycerides. TAGs are quite simple biochemicals made by plants (and animals) as energy stores. There are four basic parts to the TAG molecule: glycerol and three fatty acids (see Figure 6.2). Glycerol is a very simple molecule that links the fatty acids. Three fatty acids are chemically bonded to the glycerol. Fatty acids are fascinating substances produced by all animals, plants, bacteria and fungi. While all the vegetable oils (and fats) have the same basic building blocks, it's the structures of the fatty acids that make oils and fats so different. Fatty acids are made of chains of carbon atoms usually with even numbers of carbon atoms – 16 and 18 are the most common but higher and lower numbers are also found. Each carbon atom in the chain is bonded to two hydrogens. At one end of the fatty acid chain there is a carboxylic acid group ($-CO_2H$), which is involved in linking the fatty acid to glycerol. At the other end of the fatty acid chain there's a methyl group ($-CH_3$). Another vitally important feature of fatty acids is the double bonds that some fatty acids contain, which are formed by the removal of hydrogen atoms in fatty acid biosynthesis. Fatty acids containing double bonds are what you will have heard called unsaturated fatty acids. Polyunsaturated fatty acids contain more than one double bond and are very common in vegetable oils (and fish oils), such as linoleic acid and linolenic acid. Animal fats contain higher proportions of fatty acid without any double bonds – known as saturated fats. Different blends of these fatty acid types (unsaturated, polyunsaturated and saturated) bestow the unique properties on each type of oil. At the most basic level, the differences in the number of double bonds is what makes vegetable oils liquid and animal fats solid. But the number of double bonds also affects properties such as rancidity, with more unsaturated fatty acids being more vulnerable to going rancid; this is why processed foods contain more saturated fats.

The compositions of the fatty acids in vegetable oils and animal fats are also important for nutritional reasons; while humans (and other animals) are able to make some fatty acids, we lack the ability to make others that are needed to maintain good health. These are the polyunsaturated fatty acids, such as the omega-3 and omega-6 fatty acids that are listed and often highlighted on many food labels. These include the linoleic acid and linolenic acid mentioned above. These fatty acids that our bodies can't make, called *essential fatty acids*, are produced by many plants in high abundance, which makes vegetable oils an important source of these compounds. The triacylglycerol structure shown in Figure 6.2 contains one monounsaturated fatty acid (oleic acid) and two triunsaturated fatty acids (both linolenic acid; omega-3).

As well as fatty acids, oils contain a wide range of minor chemical components produced by the different nuts and seeds. It is these components that give different oils their unique aromas, tastes and colours. Oils vary widely in their sterol and antioxidant content, which has nutritional implications, but also offers opportunities for detecting vegetable oil adulteration. Notable antioxidants include polyphenols, Vitamin E (tocopherols), the familiar carotenoid pigments, and of course polyunsaturated fatty acids, all of which appear to be beneficial in offering protection against cardiovascular disease.

Catastrophic vegetable oil fraud

While the financial ramifications of the Salad Oil Scandal were monumental, there were no direct personal impacts on the oil consumers. By comparison, the catastrophic health consequences suffered by the Spanish people during the Toxic Oil Syndrome (TOS) scandal took the impacts of food fraud to a whole new level that made regulators sit up and take notice.

The tragic story of the TOS began on 1 May 1981 when the previously unknown syndrome began to appear in the working-class suburbs of Madrid. People were going to hospital with intense incapacitating muscle pain, breathing difficulties, headaches, rashes, itching; the diagnosis was that they were displaying an autoimmune response. Because the symptoms were similar to pneumonia, many patients were treated with antibiotics although this did nothing. The symptoms were unlike any previously encountered disease. On 10 June 1981, doctors at the Niño Jesús Children's Hospital in Madrid deduced that the probable cause of the illness was the ingestion of an illegal cooking oil sold by door-to-door salesmen as cheap olive oil. By the end of June 1981, the authorities began to seize the bottles of unlabelled cooking oil. At this point, new cases of TOS stopped appearing, but it had directly affected some 20,000 people, killing over 1,200 people in Madrid and the northwestern provinces of Spain.

So serious and unusual were the symptoms that the Spanish government approached WHO to initiate a wide-ranging international research programme, which continues to this day. The Spanish government established a centre for investigating the disease at the Institute of Health Carlos III in Madrid (Centro de Investigación para el Síndrome del Aceite Tóxico, or CISAT). The conclusion of three decades of research, published in the official WHO/CISAT report, is that TOS was associated with the ingestion of a toxic agent present in oil intended for industrial use, in this case a batch of colza oil (*colza* is Spanish for rapeseed) containing a denaturant added to render the oil unfit for human consumption. The WHO/CISAT report suggests that the denaturant added was aniline and that toxic compounds were produced during the refining process used to try and remove it. However, despite extensive investigations no compounds were found that appear to possess the extreme toxicity associated with TOS at the trace concentrations present in the oil. Only reaction products (e.g. anilides; see Appendix) between the vegetable oil triacylglycerols and added aniline have been detected.[2] The identification of the chemicals responsible for TOS remains part of the ongoing research. A paper published in the journal *Epidemiologic Reviews* in 2001[3] concluded that 'The toxic oil syndrome epidemic is an example of how even a developed country can be affected by a massive epidemic of environmental origin if failures occur in the systems that control and regulate the food supply or other consumer products.' To say that pinpointing the source of TOS has been controversial is a major understatement. Dealing with the TOS scandal drained the resources of the then newly evolving Spanish political and social medical system.

Aniline is a toxic industrial chemical with the odour of rotten fish – an effective addition to an oil that you want to ensure people don't consume. By a law passed in 1892, any cottonseed or rapeseed oil imported into Spain had to be denatured with additives such as aniline, castor oil or methylene blue. The aim was to protect Spain's olive oil industry. The Spanish government had banned the importation of rapeseed oil for human consumption to avoid any temptation to use this cheaper oil to stretch out olive oil supplies or pass it off as the more expensive oil. Yet the sale of rapeseed oil for culinary use continued, as it was an extremely lucrative business, with unwitting consumers regularly being sold various vegetable oils and mixtures as pure olive oil.

The source of the rapeseed oil in the TOS case was traced by investigators to French oil companies who were supplying rapeseed oil in France for human consumption but also producing aniline-denatured rapeseed oil for industry. The main industrial use of rapeseed oil produced in Europe was biodiesel production. Much of this industrial rapeseed oil was later found to have been diverted through Catalonia, mixed with non-denatured oils and then refined for human consumption. This became known as the Catalonian circuit. Coincidentally, in late 1980 and early 1981 the importation of aniline-denatured oil increased; subsequent investigations suggest that much of this was diverted to human consumption. A Madrid-based oil distributor called RAELCA was buying rapeseed oil for resale. The company handled various oils, including olive oil, denatured and non-denatured rapeseed oil, sunflower seed oil, grapeseed oil and oils derived from animal fat. RAELCA's process involved mixing denatured rapeseed oil with the other oils. Investigations have shown that the only toxic oil linked to a specific refinery was that associated with rapeseed oil from the ITH refinery in Seville, and the epidemic began shortly after this oil was delivered to RAELCA for retail sale. Chemical analyses confirmed the toxic oil was rapeseed oil due to high concentrations of the diagnostic compound brassicasterol, which is a phytosterol (a plant equivalent of the animal sterol cholesterol) produced by rapeseed (we discuss sterols more in Chapter 6 and provide structures in Fig. 6.2). The repercussions of this epidemic were widespread, resulting in many of the implicated oil refiners and distributors being convicted and jailed.

There are theorists who firmly believe the TOS was an elaborate diversion intended to cover up a potentially more damaging source of toxin. The most prominent of these ideas was that the epidemic was the result of tomatoes and possibly other vegetables grown in Andalusia contaminated with organophosphate pesticides, particularly isofenphos and fenamiphos. Although this generated considerable media publicity, the idea appeared not to hold up because the symptoms of TOS differed from known organophosphate toxicity symptoms.

The oil identity crisis

Fraud in the vegetable oil business could not be easier as it is simply a matter of mixing two liquids. The physical similarities of different oils makes a sublime situation for the cheats; it's somewhat akin to the situation in the wine trade, but we'll come to that in Chapter 8.

In their raw states, the colours of vegetable oils vary from virtually colourless to yellow to orange to green. However, appearances can be deceptive, especially when methods exist for adding colour and taking colour away. The odours of oils vary, but aromas and flavouring can also be added and removed. So the technically adept fraudster can make one oil look like another. We have no idea just how prevalent vegetable oil fraud actually is, but it is rampant enough among commonly tested oils to suggest that there is an underlying chronic fraud.

Olive oil is an especially popular commodity on account of its desirable flavour and perceived health benefits. Like some other speciality oils, it is produced in relatively low volumes and commands a high price. This makes it especially prone to adulteration, despite international efforts to regulate it. In contrast to these premium oils, others are produced cheaply in much higher volumes. So the temptation is just too much: either make the cheap oil smell and taste like the expensive oil or add significant amounts of the cheap oil to pad out the expensive oil. It's all about increasing the profit margin by selling a cheaper product and leading the consumer to think it's the more expensive product. This is deeply immoral and damaging to the producers of the genuine article, but until there is a disaster on the scale of the TOS vegetable oil fraud, it attracts surprisingly little attention.

The massive production volumes of vegetable oil make it impractical to test every batch. Additionally, the natural variation in the physical and chemical properties issue raises its ugly head again; the properties of even the same type of oil vary over considerable ranges, making it difficult to define acceptance criteria and standards. And it is within these ranges of variability that fraudsters do their business.

Cracking vegetable oil fraud

Discerning the differences between the different oils should be a matter of looking for differences in the main constituents – fatty acids. However, the fatty acid compositions of the vegetable oils listed in the Codex Alimentarius show that these oils are mainly composed of the same fatty acids. And the mixtures of these fatty acids vary naturally over sufficient ranges to give the cheats plenty of scope to mix two very different oils with similar fatty acid profiles. For example, if we compare the three major fatty acids in olive oil, the natural range in the proportions of each fatty acid overlap with other oils. And it doesn't matter which olive oil we use for comparison because virgin, refined and olive pomace all have the same fatty acid compositions. Indeed, there is so much scope for mixing oils together that we can only guess at the scale of the fraud.

In trying to detect vegetable oil fraud we have to confront two questions: (i) does the oil in question contain an undeclared oil? And (ii) if there is an undeclared oil, how much of it is present in the bottle? The first question is often easier to resolve than the latter. Progress in answering both these questions has been made with maize or corn oil, with a test being developed that allows its purity to be accurately determined using a novel approach. The international production of maize oil is currently more than three million tonnes; that's nearly four billion one-litre bottles. Maize oil is popular because of its good shelf life, pleasant flavour, stability and healthy fatty acid profile.

Maize oil doesn't carry the premium price that some oils do, but it is still twice as expensive as products labelled 'vegetable oil' (which are primarily rapeseed in Europe and soy in North America); it's also produced in high volume. This makes maize oil a prime target for fraudulent adulteration with cheaper vegetable oil. The first hints of extensive maize oil adulteration came in the 1990s as a result of investigations by the then UK Ministry of Agriculture, Fisheries and Food. Some 291 edible oils were sampled from retailers and submitted to fatty acid, phytosterol and tocopherol analyses. Phytosterols are found in all plants with some of the structures characteristic of particular plants. Brassicasterol, which we mentioned previously, is characteristic of rapeseed oil, for example (see Figure 6.3). Likewise, tocopherols, the compounds that make up Vitamin E, also possess compositions specific to different oils. The analyses found that around 80 per cent of the collected oils were correctly labelled. However, of the 79 maize (corn) oils analysed, 35 per cent were deemed to contain an undeclared oil. The sterol and tocopherol analyses indicated that the most commonly added oil was rapeseed oil.

The results of the survey suggested that the adulteration of vegetable oils was surprisingly commonplace. However, the existing tests had their shortcomings. The detection and quantification of adulterant oils in maize oil presented a particular problem. The fatty acid composition of maize oil has an unusually high natural variability, such that blends of oils may readily be prepared with fatty acid compositions lying within the range expected for pure maize oils. Furthermore, the unusually high sterol and tocopherol contents of maize oil also help to mask the presence of added adulterant oils.

Another test was needed. In the early 1990s Barry Rossell, working at the Leatherhead Food Research Association (LHFRA), UK, recognised that maize came from the group of plants that use the C_4 photosynthetic pathway, which you may remember from our honey case study in the previous chapter. All other major vegetable oil-producing plants use the C_3 pathway. Rossell had the idea of using this biochemical difference as the basis of a new test. He collected a range of oils together and submitted pure oils and mixtures, prepared to mimic adulterated oils, for carbon isotope analysis. This was a very new approach to detecting food adulteration at that time. His results were extremely promising, showing that as little as 10 per cent of a C_3 vegetable oil could be detected in maize oil. However, even 10 per cent adulteration is a lot of cheaper oil being added. And when you think that a typical UK supermarket price per litre for maize oil is twice that of rapeseed oil, this constitutes significant economic fraud globally, even if there aren't likely to be any significant health consequences. The detection limits needed to be improved.

Around the same time, in the early 1990s, a new type of isotope ratio mass spectrometer became commercially available – a gas chromatograph-combustion-isotope ratio mass spectrometer (GC-C-IRMS). The idea for the technique came from the organic geochemist John Hayes working in the US. One of the first commercial instruments in the UK was acquired by a group in the School of Chemistry at the University of Bristol. One of us (Richard) joined this group shortly thereafter and was looking to put this new technology to good use. Recalling a lecture given six months earlier by Barry Rossell, Richard saw an opportunity to improve the maize oil stable carbon isotope test by looking at the carbon isotopes in the individual fatty acids rather than the whole oil. Rossell was contacted and a pilot study was initiated by Richard's MSc student Simon Woodbury. A handful of maize and rapeseed oils were analysed and the new technique could detect down to five per cent of rapeseed oil in maize oil – a huge improvement over the existing ten per cent.

So encouraging were these results that the LHFRA funded Woodbury to undertake a PhD to further develop this new technique. He put together a new database of more than 150 vegetable oil fatty acid compositions, together with their individual carbon isotope values. Since oil adulteration was so rife, commercial oils couldn't be trusted to build this new database, so Woodbury undertook his own painstaking extractions of the actual seeds.[4, 5] The work established a precedent for how this type of compound-specific isotope work had to be done. It also showed how carbon isotope values for minor components of the oils, namely sterols and tocopherols, could be used in conjunction with those obtained for the fatty acids to significantly improve the threshold of detection for adulteration of maize oils.[6] After presenting the paper on the new technique to an oil producer meeting it was interesting to see how "pure" the maize oils on the supermarket shelves became!

The FSA used this technique in a follow-up investigation in 2001. They tested 61 samples and found none were adulterated. Perhaps the most encouraging aspect of this work is the fact that despite no changes in regulation, an improvement in the detection technique had on its own led to a reduction in fraud. The technique was ultimately incorporated into the international standard in the Codex Alimentarius and for once the detection technique was ahead in the scientific arms race!

Olive oil: a likely victim

The only benefit of any food fraud scandal is that it creates an opportunity to crack open the food production and distribution systems that are elusive to most of us, find the vulnerabilities, and try to correct them. The Spanish Toxic Oil scandal was evidence that olive oil fraud was probably commonplace. Fraud in olive oil derives from the coalescence of four major elements:

1. *Consumer demand for high quality olive oil* and the misguided belief that somehow an oil advertised as cheap can also be high grade. The same principle applies to all things – you get what you pay for.

2. *The production of nine different grades of olive oil,* including four different grades of virgin olive oil, two grades of olive oil and three grades of pomace olive oil. To quote the International Olive Council (IOC),

> Virgin *olive oils are produced by the use of mechanical means only, with no chemical treatment. The term virgin oil includes all grades of virgin olive oil, including: Extra Virgin, Virgin, Ordinary Virgin and Lampante Virgin olive oil products. Lampante virgin oil is extracted by virgin (mechanical) methods but is unsuitable for human consumption without further refining;* lampante *is Italian for 'lamp' and refers to the earlier use of such oil for burning in lamps. This oil can be used for industrial purposes, or refined to make it edible.* Refined Olive Oil *is the olive oil obtained from any grade of virgin olive oil by refining methods which do not lead to alterations in the initial triacylglycerol composition. The refining process removes colour, odour and flavour from the olive oil, and leaves behind a very pure form of olive oil that is tasteless, colourless and odourless and extremely low in free fatty acids. Olive oils sold as the grades* extra-virgin olive oil *and* virgin olive oil *therefore cannot contain any refined oil.* Crude Olive Pomace Oil *is the oil obtained by treating olive pomace (the leftover paste after the pressing of olives for virgin olive oils) with solvents or other physical treatments, to the exclusion*

of oils obtained by re-esterification processes and of any mixture with oils of other kinds. It is then further refined into Refined Olive Pomace Oil and once re-blended with virgin olive oils for taste, is then known as Olive Pomace Oil.

3. *The huge scale of production of cheaper oils,* such as sunflower oil, provides an obvious source of potential cheap additive oils for the would-be fraudsters. Some of these oils have similar chemical compositions to olive oil, which helps to mask the additions unless more sophisticated tests are used. In other words, the fatty acid compositions of the two oils may be very similar, but they differ in their sterols and tocopherols.

4. *The lack of simple tests for the different grades of olive oil.* Consulting the Codex Alimentarius, EU marketing standards and the IOC testing methods reveals the problem. One of the primary ways to assess the purity of olive oil is an analysis of its organoleptic characteristics – that is, its smell and taste. Aficionados of olive oil argue that few of us would be able to recognise a high quality olive oil if it was put in front of us. This has been a challenge for the relatively new California olive oil industry. US consumers are not used to consuming fresh olive oil and so perceive this fresh flavour as being 'off'. Most of us do not live in olive oil-producing regions so our exposure to fresh olive oil will be either rather limited or actually non-existent. Assessing the purity of olive oil from its smell and taste sounds like an art but in fact it's an amazing piece of science. Sensory assessments rely on trained testers to objectively describe the characteristics of an oil. Their descriptions are then statistically analysed to ensure they are due to differences in the product

being tested rather than differences between the testers. The conditions published by the IOC in February 2015 for undertaking such tests are very strict. They dictate how tasters should be selected and trained and how the testing should be performed. They even dictate how the room should be arranged and the shape and size of the glass used for testing the predetermined volume of oil. It is also recommended that tests are performed between 10 a.m. and 12 noon as it has been shown that smell and taste sensitivity increases at that time of the day. The testers are trained to recognise the positive attributes for fresh olive oil, which include fruity, bitter and pungent sensations, varying in intensity depending on the variety and ripeness of the fruits. Negative flavours include: heated or burnt, hay-wood, rough, greasy, vegetable water, brine, metallic, esparto, grubby and cucumber, characterising substandard olives or oil resulting from inappropriate quality, storage and/or processing of the fruit. The test is the cornerstone of olive oil authenticity testing and as we will see later it was crucial in revealing widespread olive oil fraud.

If the normal commercial spectrum of grades is not enough encouragement for the cheats, they must really begin to rub their grubby hands together when further opportunities are added through the olive oil connoisseur culture. Additional 'grades' have been introduced based on renowned regions and preferred local producers. For example, reputedly the most expensive olive oil in the world is Lambda, which is produced by Speiron Co. in Greece. The oil is labelled as an 'Ultra Premium Extra Virgin Olive Oil' and is made from Koroneiki olives, which are harvested by hand and cold pressed in order to avoid heat altering the flavour, aroma and nutritional value. The oil is packed by hand and sold in an attractive 500ml (17 fl oz) bottle. Lambda costs £34.50 (US$54) per bottle, but if you want something a little more special your bottle of Lambda is offered in a gift box at £128 (US$200). But why stop there? For true olive oil connoisseurs you can secure a bottle of Lambda in a hand-crafted case with two 18k gold plates, one of which, along with the bottle, bears the owner's name; all for an unbelievable cost of £9,433 (US$14,698). We're not saying there's anything wrong with anyone enjoying the very best food, but this type of exclusivity and ultra-high price introduces further opportunities for fraud.

The olive oil hits the fan

Motivated by the dominance of European olive oil on US supermarket shelves and concerns over its authenticity, the University of California Davis (UC Davis) Olive Centre initiated a study in 2010 of how olive oils are tested.[7] The study, led by Dr Edwin Frankel, was conducted in collaboration with the Australian Oils Research Laboratory. Together they investigated 186 extra virgin olive oils purchased from retail outlets in California. The samples included both imported and locally produced brands. The two laboratories had independent sensory panels in Australia and California evaluate the oils, using methods recommended by the IOC. Remarkably, 73 per cent of the oils failed the IOC organoleptic sensory standard for extra virgin olive oil. The failed samples had objectionable descriptors, such as rancid and fusty. These same oils were subjected to the IOC's recommended chemical analyses, which included fatty acid profiles. Oils that had failed two IOC-accredited sensory panels worryingly passed the chemical analyses. The team then applied more sophisticated tests. The German Society for Fat Science (DGF) had developed tests targeting other components in the oil, such as diacylglycerols (DAGs) (like a triacylglycerol, but with only two fatty acid chains) and pyropheophytin (a compound formed in the degradation of chlorophyll and commonly called PPP). When Frankel and his team applied these methods to the same oil samples, 70 per cent of the oils from the five top-selling imported Italian brands failed the DAG test and 50 per cent failed the PPP test. The study showed not only that the majority of top-selling imported brands of 'extra virgin' olive oils sold in the US were failing sensory tests, but that the IOC's recommended chemical analyses weren't sufficient. The oils were failing due to oxidation by exposure to elevated temperatures, light and/ or ageing, or adulteration with cheaper oils, or they were poorly made from low quality olives – or a combination of these factors.

The results of the UC Davis study were shocking. But then came Tom Mueller's explosive book *Extra Virginity*, published in late 2011. Mueller provides a compelling exposé of the chaotic world of the olive oil trade in which the conflict is emphasised between farmers and big business. The bottom line was that it appeared Italy was exporting more olive oil than it was producing, and the mislabelling exposed by the UC Davis investigation, and others, was rife. It was probably no coincidence that in January 2012 the EU codified a decade's-worth of amendments to the olive oil marketing standards introduced in 2002. In particular, the new version of the regulations (EU No 29/2012) clarified origin labelling on olive oils in an attempt to offer a degree of control and reassurance to consumers in the wake of mislabelling incidents concerning products 'Made in Italy'. However, with the cat out of the bag, it was pretty clear that it would take more than a few new regulations to ensure consumers were buying the product depicted on the label. The apparent deficiencies in the established tests revealed by the UC Davis study required a major rethink in the methods used to authenticate extra virgin olive oil.

Faced with new EU labelling laws, new growing regions emerging into the market and an industry ripe for fraud, what was the way forward in terms of analysis? In June 2013, olive oil analysis experts from all around the world, including Frankel from UC Davis, gathered in Madrid to discuss olive oil fraud. The workshop considered the current status of the olive oil market and the trade standards for olive oil globally. Participants undertook a detailed review of the methods available for detecting frauds of various types and considered the magnitude of challenges in detecting olive oil fraud. They accepted that sensory testing was clearly important, but recognised the need to link panel testing to a better understanding of the chemistry behind the organoleptic qualities of olive oil. More research was needed to know how the different production methods affected the chemistry of the oil and how these different chemistries were correlated with the results of the sensory testing. Logically, the connection was made between the assessment of organoleptic properties and possibilities for using new technologies.

Electronic noses and tongues were seen as being an area ripe for development. These devices contain sensors that try to mimic the sensory capacity of the human nose and tongue, but with more objectivity. The so-called *e-noses* are designed to detect volatile compounds while the *e-tongues* detect substances in solution. The idea is that electronic signals are produced by the sensor arrays, which are then analysed using multivariate statistics to reveal patterns in the data that allow different samples to be compared. While the goal of removing the subjectivity from organoleptic panel testing is laudable, the challenge of ever achieving the same sensitivity and specificity is very considerable. More conventional analytical chemistry approaches were also highlighted as an area for development, particularly methods that would reveal more information about the volatile organic compounds that lie at the heart of the organoleptic panel tests. With the expansion in production regions and introduction of new olive varieties, DNA-based methods are also likely to have a role in recognising mixtures of oil. However, significant work lies ahead before such methods become an established part of the armoury of techniques for use in the battle against the fraudsters.

A perfect storm

The US Pharmacopeia Food Fraud Database currently lists more than 300 reports, the majority scholarly articles, relating to olive oil fraud. The most common type of fraud is the misrepresentation of the type of olive oil, either its grade or the country of origin. There are also numerous instances of olive oil being mixed with other cheaper oils such as soy, sunflower, rapeseed and corn, with less common additions being grape, peanut, cotton, mustard, sesame, palm, walnut and almond oils. The nut oil adulteration is a particular concern as this brings with it the risk of allergic responses from consumers.

While the Madrid workshop was a step towards recognising what tools were absent in the fraud fighting toolbox, the regulatory authorities in Europe still needed to be spurred into action. The trigger came in the form of the 'perfect storm' in 2014; bad weather and pest infestations occurred simultaneously across southern Europe, causing widespread damage to olive crops in the region. The inevitable olive shortages exerted considerable pressures on supply chains. The reduced supply meant the demand for olive oil could not be met, with the result that fraudulent activity increased. The events of 2014 have been called the *annus horribilis* for olive oil fraud, particularly in Italy. Italian olive oil supplies were massively reduced but because the 'Made in Italy' label earns a premium price, the inevitable result was fraud: many of the olive oil products claiming to be Italian were not.

Alert to this, Italy's inspection and regulatory body for the protection of quality and fraud prevention of food products, known as ICQRF, carried out a multi-agency campaign to track the movements of vegetable oil from Italian ports to production plants, distributors and commercial outlets. They checked 4,114 operators and considered 452 of these to be 'irregular'. They checked 6,004 products and found that 569 (9 per cent) failed to meet regulatory standards. They took 1,195 olive oil samples and had them extensively analysed by official European panels and found that 66 (6 per cent) were 'irregular'. Some 140 administrative penalties were levied and 122 seizures were made with a total value of €9.8 million (£7 million, $11 million). Similar concerted actions were revealing analogous widespread fraud in other regions of the EU. To his great credit, on 21 January 2015 the Italian Minister of Agriculture, Maurizio Martina, organised a meeting to reflect on the status of the Italian olive oil industry. The meeting involved regulators and key stakeholders in the Italian olive oil supply chain. He defined a long-term strategy against counterfeit 'Made in Italy' oil that strengthened counterfeiting interventions, including providing financial support to Italian producers in times of strife such as the *annus horribilis* of 2014.

In response to global evidence of vegetable oil fraud, plus a European Parliament report that highlighted olive oil as one of the products (together with fish and organic foods) most prone to food fraud, the EU announced funding in 2014 to advance the science that might help in catching these slippery criminals. The call for submissions came under their Horizon 2020 scheme and the budget was €5 million (£4.4 million, US$6.7 million) to tackle olive oil authentication. The announcement acknowledged the position of the EU as the world's largest producer, consumer and exporter of olive oil. And it was calling for proposals for research that would further the development, validation and harmonisation of 'analytical methods and quality parameters that specifically address technical authenticity issues'.[8] In particular, they wanted to address the blending of extra virgin or virgin olive oil with lesser quality oils (olive or otherwise).

The announcement was an indictment of the unsatisfactory state of the olive trade and testing practices globally. To quote Alex Renton's review of Mueller's book in the *Guardian* newspaper:

> ... *you could tell the same story of almost any artisan's product we put in our mouths, from bacon to cheddar cheese or smoked salmon. Industrial production techniques and the supermarket's tendency to strip out quality in order to give 'value' will debase any foodstuff once it becomes popular to the point where the producer has to abuse his animals, sin against tradition or commit fraud in order to stay afloat.*

Strong words indeed, but we will be providing an abundance of examples of exactly these sorts of practices in subsequent chapters.

So what of olive oil? The EU research is only just starting and it will be years before the results have any genuine impact in the marketplace. But as the Italian initiative has shown, the regulators can have an immediate effect. If the production of extra virgin olive oil is more closely controlled, there will be knock-on consequences: you will have to pay more for the best authentic products – a price that better reflects the cost of producing it. One of the most critical pieces of guidance when contemplating a purchase of extra virgin olive oil is that if it's cheap it's probably not the real thing. Beyond this, we are in the hands of the suppliers and their internal controls to regulate their supply chains and ensure our olive oil is what it says on the bottle.

CHAPTER FOUR
Hake Today, Cod Tomorrow

In May 2003, the first Canadian-born beef cow tested positive for bovine spongiform encephalopathy (BSE), commonly known as mad cow disease. Eating diseased cow products is linked with variant Creutzfeldt-Jakob disease (vCJD) – a rare but fatal disease in humans. In both BSE and vCJD, normal proteins that are found on the surface of neurones start to change their shape so they are no longer functional (misfold). These abnormal infectious little proteins, called prions, cause misfolding in other proteins and then they begin to gather together. These aggregations eventually create a type of fibre that's commonly seen in other neurological diseases such as Alzheimer's, Huntington's and Parkinson's. Sometime between the first prions getting together and the formation of the fibre, the original neurone is killed. As the neurones die, holes form in the brain of the infected animal and ultimately lead to its death. Unlike bacteria and viruses, prions can't be broken down with normal cooking. In 1986, when England made its first diagnosis of BSE, five million cattle were destroyed to stop the spread of BSE, but not before one million untested cattle made it into the food supply. The first person to develop symptoms of vCJD became ill in the UK in 1994; there is a long incubation time while the misfolded prions reach any symptomatic critical mass. Deaths peaked in 2000 when 28 people died of the disease. There have been around 220 cases of vCJD worldwide to date, but it is estimated that there are approximately 15,000 citizens in the UK who are potentially incubating the disease, so this story is not yet over.

What does mad cow disease have to do with food fraud, and why are we talking about beef at the beginning of a chapter on fish, you ask? In 2003, when that single Canadian-born cow was found to have BSE, a fish processor in Ucluelet, British Columbia (BC) was forced to shut down early in the season. The processing plant turned hake into surimi – the gelatinous paste that is flavoured, coloured and shaped into flakes, sticks and other wonderful shapes and used in fish cakes, fish burgers, fish balls and the imitation crab meat so commonly found in California rolls. Surimi is a clever way of converting a cheaper white fish into likenesses of higher-end products, such as crab, eel and lobster. As part of the process, beef blood plasma – the clear part of the blood that is left over when you remove all the cells and platelets – is added to the fish paste to help it form a gel (these days egg and something called transglutaminase, which we discuss in the next chapter, are more commonly used). Blood plasma products are widely used as gel enhancers in the food industry. The BC plant lost business because nobody wanted a beef product that was made in Canada, despite the fact that the Canadian processors were obtaining their beef plasma from the mad-cow-free US.

Just to be clear, we are not saying that this was in any way a form of food fraud. It was probably written right on the label, most likely as bovine blood plasma or perhaps fibrinogen or thrombin. It is, however, an example of how our food is crossing taxonomic boundaries – horse in our beef, beef in our fish and so on. This creates complications (and opportunities) in terms of food authenticity testing, but it also means that when a food safety issue such as BSE rears its ugly head, understanding its reach within our food supply is daunting even in a world where the food we buy is labelled correctly. It gives one little hope that we can track it down in a world of substitutions and scams – two things very familiar when it comes to seafood.

Susceptibility to pseudonyms

Like any other food, seafood – fin fish and shellfish – has been prone to fraud throughout history. Recall the painted gills in the night markets of London in Chapter 1? The most commonly talked about fraud within the industry today, however, is mislabelling – mostly in terms of the species name and/or where it was caught. Owing to a combination of factors, seafood, particularly fish, is more prone to mislabelling than any other protein source in our food supply and there are a number of reasons why this might be the case.

First, our desire for fish products is on the rise. Global fish consumption has been growing at a rate of 3.6 per cent per year since 1961 – faster than can be accounted for through population expansion alone. Global per capita fish consumption has grown from 9.9kg (21.8lb) per person in the 1960s to 19.2kg (42.3lb) per person in 2012. Campaigns on the health benefits of eating fish have been widespread, leading many consumers to incorporate more seafood-based protein into their diet. And with more sophisticated processing, better storage techniques and endless capacity to move food around the planet quickly, fish and other seafood products are now accessible to regions that would not have had easy access to these products historically. Fort McMurray, Canada, for example, is over 1,600 kilometres (1,000 miles) by road from the nearest ocean, yet the town of 77,000 people supports at least four sushi restaurants.

Second, there are supply challenges inherent to the industry. Fisheries represent the last remaining commercial-scale harvest of animals from the wild (though the balance is shifting now that half of global production is from aquaculture). In the past decade we have consistently captured about 90 million tonnes of fish from the ocean and inland waterways each year. Management of this resource is rather difficult though, as it's really a matter of trying to count animals we can't see in an environment we can't control. As a result, many fish stocks have been depleted, mostly because (to use the words of Professor Daniel Pauly in the film *End of the Line*) 'we ate them'.

When fisheries' stocks start to dwindle, managers are obliged to implement strict harvesting restrictions in order to help the stocks recover. This, by default, creates a premium product as scarce species are generally worth more on the market. In the 1980s, US fisheries managers put in place a number of regulations to help recover stocks of red snapper, which the US FDA has stated is the legally accepted common name for the species *Lutjanus campechanus*. Among other restrictions, there are limits on the number of fish that can be caught each year, so those that are caught sell for a good price. In 2011, red snapper averaged a value of about US\$7.04 (£4.50) per kg (US\$3.20 per lb). That same year, Labrador or Acadian redfish, *Sebastes fasciatus,* which has been substituted for red snapper, was worth a measly US\$0.56 (£0.36) per kg (US\$0.25 per lb). It is difficult, if not impossible, to differentiate the two species once they are fillets and the economic incentives to mislabel are substantial.

As mentioned already in Chapter 1, climate change is unlikely to improve the reliability of fish stocks. The range of some species may shift. Some species may thrive while others fail. There have also been predictions that in the future severe storms may become more frequent. Fishermen already contend with tides and weather, so frequent storms and big waves could be an additional challenge the industry faces in the future.

Perhaps more than anything else, however, it is globalisation of the industry that has made it particularly vulnerable to fraud. The seafood trade expanded globally in the early 1990s and the US's National Seafood Inspection Laboratory (NSIL) began routinely examining seafood products. They reported that between 1988 and 1997, their tests confirmed that 37 per cent of fish species and 13 per cent of other seafoods were mislabelled; as many as 80 per cent of red snappers tested were mislabelled.[1] This was some of the first evidence of the prevalence of mislabelling.

In 2013, the US imported just under 2.5 million tonnes of edible fishery products worth approximately US$18 billion (£11.3 billion). This is nearly double what the US imported 20 years earlier, in 1993 (1.3 million tonnes). The UK, with about 20 per cent of the US's population, imported approximately 739,000 tonnes, valued at £2.6 billion (US$3.9 billion). While importing such vast quantities of fish products helps meet demand and introduces new and exciting products into the market, it also creates opportunities for labelling mistakes and intentional fraud.

As we have seen over and over again, vulnerability to fraud is introduced with every link in the food chain; fish caught and sold in one place are being shipped around the globe for processing, introducing some very distant links. Fish and seafood caught in Alaska, for example, are sent to China for processing at one-fifth to one-tenth the cost of processing in the US. The fish are filleted and the crabs are de-shelled, then they're packaged and given a 'Product of China' label before being shipped back for the US market. The Codex Alimentarius states that if a food has been processed in a way that changes the nature of the product, the origin of that food becomes the country in which it was processed. Russian sockeye salmon that are processed in BC, Canada, become a product of Canada and are labelled as BC salmon. It's perfectly legal, but very misleading. About 90 per cent of the 104 million kilograms (230 million pounds) of squid caught in California each year is sent to China for processing before ending up back in the US for sale – a 19,000-kilometre (12,000-mile) round trip. Despite fuel costs, it is still cheaper to have seafood processed overseas, and cost is critical for an industry trying to keep afloat among cheaper imported and farmed seafood products.

Fish are so well travelled, in fact, that they are one of the most traded food commodities in the world. And, as anyone who has travelled knows, there are challenges associated with operating in different countries among unique cultures, foreign languages and unfamiliar regulations. Then add to this the complication of common names. Many species are given common names to try and simplify things – it's undeniable that clownfish is far more memorable than *Amphiprion ocellaris*. However, a single species can accumulate many names over time, which can complicate matters. Let's take Atlantic cod, species name *Gadus morhua*, as an example. It has nearly 200 known names. In the English language alone this species has at least 58 names, 56 of which are used in Canada. We couldn't possibly provide all the names, but some of our favourites include bastard, blackberry fish, duffy, foxy, tom-cod, grog fish, hen, loader, old soaker, pea, snubby, split and swallow tail. It would be equivalent to you showing up at border control with 200 passports each bearing a different name. A private interrogation room would probably figure in your near future.

Some of the name confusion is generated by marketing campaigns aimed at making species names somewhat more palatable, or at least marketable. For example, Marks & Spencer received permission to rename *Glyptocephalus cynoglossus* from witch flounder to Torbay sole (a nice local-sounding name). Between 1973 and 1981 the US National Marine Fisheries Service (NMFS) spent half a million US dollars (£315,000) looking at underutilised species to determine which ones could benefit from a name change. It was during this period that slimeheads (*Hoplostethus atlanticus*) became orange roughy.

To try and overcome the challenges of multiple names, many countries have issued lists that provide the legally accepted market names for each species sold in that country. Canada's fish list states that out of the 56 English names for cod used in Canada, only cod and Atlantic cod are acceptable market names. But differences are bound to exist between countries. Basa, for example, can refer to the catfish species *Pangasius bocourti* or *Pangasius hypophthalmus* in Canada, while the US FDA states that this common name can only refer to *P. bocourti*. In the UK, basa can refer to any of the 21 species within the genus *Pangasius*. There are over 60 species of fish on the FDA's list that can be sold as grouper, while the UK simply states that any fish from the two genera *Epinephelus* and *Mycteroperca* can be labelled as grouper — a combined total of over 100 species.

Consumer demand for more non-fuss meals also encourages opportunities for mislabelling. Many of us don't want to come home from work and start filleting fish, cleaning squid or digging crab meat out of the shell before we cook our evening meal. In general, consumers are looking for prepared portions of fish that we can cook up in a matter of minutes. In response, supermarkets place such specific requirements in terms of how the fish looks and what the portions weigh that they need to be hand processed. This is a round-trip ticket to Asia. And it is not only consumers that are looking for a prepared product — restaurants are too. Many regions have very rightly established regulations to minimise food waste and recycle it in composting facilities rather than having it end up in landfill. This often includes a levy or fee based on the weight of food waste produced by the establishment each week. It therefore doesn't pay to bring in whole seafood products that kitchen staff have to take the time to prepare, and that also contribute significantly to their weekly waste. Commercial kitchens with savvy chefs are already doing what they can to reduce food waste as every scrap in the bin is money lost. Buying pre-prepared fillets and portions of fish is simply an extension of this efficiency. It has meant, unfortunately, that many chefs are losing some of the most basic skills of preparing seafood for cooking and that the entrees may be better travelled than many of the restaurant's customers.

Should we then be surprised that as portions of white, boneless, skinless, nondescript fish travel around the world, their names (of which they have many) and where they were caught have a tendency to change, whether accidentally or intentionally? Perhaps not, but how often it happens may be surprising. And with heads, skin and most other recognisable features removed, we need to turn to DNA-based methods to determine whether the cod goes by any other name.

Dealing with diversity

Before we get into the nitty-gritty of DNA analyses, however, we feel obliged to lay out just how monumental a task it is to identify the species and origins of fish. There are over 32,000 known species of fish. They are, by far, the most diverse group of vertebrates on the planet, making up more than half of the 62,000 (give or take) identified vertebrate species. Estimates available on the number of species used for food globally range from 900[2] to 20,000.[3] The FDA's seafood list contains over 1,800 records of seafood species encountered in US markets. This would include species that don't have a targeted fishery, but are caught as by-catch and then sold. The point is, there are a lot of species.

Compare the more than 1,800 fish and seafood species listed in the US with the number of species of meat sold in the US – namely, beef, pork, chicken, turkey and, to a much lesser extent, lamb, mutton and goat. There are about 13 breeds of cattle reared for meat in the US and they have all been bred from one of three species: *Bos taurus, B. indicus* and bison or hybrids thereof. There are hundreds of breeds of pig, but all of them come from just one species – the Eurasian wild boar, *Sus scrofa*. There are 34 different breeds of chicken reared for meat, all of which belong to the single species *Gallus gallus domesticus*. There are eight different varieties of turkey recognised by the American Poultry Association, all belonging to the domesticated turkey species *Meleagris gallopavo*. When it comes to terrestrial protein, we're only really dealing with a handful of species and this helps simplify testing.

To try and differentiate the thousands of fish species out there, protein-based techniques have traditionally been used. Each species has a unique protein profile and unknown species can be identified by comparing their profile with known species samples. Isoelectric focusing is the technique frequently used to do this. The principle is that a molecule, such as a protein, will lose its electric charge at a particular pH. This is known as the molecule's isoelectric point and it can be used to separate out different proteins. An unknown sample is prepared by mixing tissue from the fresh or frozen fish with water. The water-soluble proteins dissolve out into the liquid and then it's centrifuged to get rid of any bits that didn't dissolve. The sample is added to one end of a gel (think very thick jelly in a thin layer on a plate) that has a pH gradient from one end to the other. A current is applied, which causes the charged proteins to travel through the gel until they reach a pH where they are no longer charged. Each type of protein will stop at a unique point, creating very distinct band patterns that are relatively unique to the species and can then be compared with a known sample for identification.

It's not unlike taking an unknown, unlabelled bag of Lego pieces and running them through a series of sieves to separate the pieces by shape. You would then be able to compare what you've got with known Lego kits and could say that the unlabelled bag must be kit no. 31002, the Super Racer, because it contains 29 different types of Lego pieces, four single white pieces, two double yellow pieces, and so on.

Barcoding life

However, just as Lego builders are thwarted by the common vacuum, protein analysis is thwarted by the common oven. While protein-based techniques are quick, simple and cost effective, their utility in the food industry is somewhat limited as proteins are denatured during cooking and many other standard processing methods. They're also limited because proteins are expressed differently in different tissues; the protein profile produced from a sample of Pacific halibut skin, for example, may not be the same as that produced from a sample of its muscle tissue. Some protein-based methods may have difficulty differentiating between closely related species as well. It is for these reasons that analysts are going straight to the source, to the blueprint that codes the proteins and, indeed, life itself – DNA. It is not as easily degraded by heat and it's found in almost all cells of living organisms (an exception being red blood cells). DNA-based methods have also become far more affordable in recent years.

Many people within the scientific community point to a brief communication in 2004 in the journal *Nature* as the first use of DNA-based methods to reveal the prevalence of fish mislabelling. Peter Marko, then at the University of North Carolina and now at the University of Hawaii, and his team used DNA analysis and found three-quarters of fish being sold in the US as red snapper were species other than *Lutjanus campechanus*.[4] Marko and his team analysed fish purchased from nine vendors across eight states and compared DNA sequences from these samples with those found in the open access sequence database GenBank. They found that 77 per cent were other species. With the margin of error associated with the technique, anywhere from 60 to 94 per cent of the samples could have been mislabelled. Five of the mislabelled species were other forms of Atlantic snapper and two of the samples were crimson snappers, which are from the Indo-West Pacific. A number of species couldn't be identified because they were either from other regions of the world or were too rare to be included in the GenBank database.

It is possible that the mislabelling of the other Atlantic snapper species could have been genuine misidentification on the boat as these species are likely to have been harvested at the same time. This, unfortunately, can lead to inaccurate catch statistics – overestimating the harvest (and by default the abundance) of red snapper, while underestimating the misidentified species. Those species caught from other areas of the world, however, are likely to have been mislabelled at some point after they came off the boat because it seems unlikely that an Indonesian fisherman would misidentify a crimson snapper as an Atlantic species. Regardless of where along the supply chain the mislabelling happened, Marko's conclusion was that this rampant mislabelling was distorting consumer perceptions about the availability of red snapper.

The year before Marko published his paper on red snapper, Professor Paul Hebert from the University of Guelph, Canada, was proposing DNA barcoding as a method of identifying species. You will recall this method from our honey example in Chapter 2. However, unlike for plants, the DNA segment Hebert and his colleagues at Guelph targeted for animal barcodes was about 650 base pairs long and codes the protein cytochrome c oxidase subunit 1 – more easily referred to as CO1. It is one of the subunits that make up the enzyme cytochrome c oxidase, which is found in the powerhouse of the cell – the mitochondrion. It plays a critical role in the cell's energy production.

Targeting mitochondrial DNA has several advantages. First, there is more material to work with as mitochondrial DNA is far more abundant than nuclear DNA. Most cells have only a single nucleus, but they may have hundreds of mitochondria. In fact, a human liver cell may have as many as 2,000 mitochondria. In terms of sampling, this means that it's easier to recover DNA from degraded material. Second, mitochondrial DNA has a higher mutation rate compared with nuclear DNA, which means there is a better chance of finding differences between species. Third, mitochondrial DNA is only inherited from one parent, which simplifies matters in terms of sequencing the DNA. As mammals, we have each received equal contributions of nuclear DNA from our mother's egg and from our father's sperm. We therefore have two complete sets of chromosomes – two different versions of a gene – which makes us diploid. However, the mitochondria that enter the egg from the sperm are destroyed early in development, leaving us with only the mitochondria we inherited from our mothers and therefore only one copy of mitochondrial DNA (haploid).

Hebert and his colleagues put this segment of DNA to the test and examined the COI sequences in GenBank of more than 26,000 animals from 11 broad taxonomic groupings (e.g. worms, crustaceans, beetles, flies, wasps and bees, butterflies, chordates, jellyfish and molluscs).[5] They found that with the exception of the jellyfish and corals (cnidarians), they could clearly tell species apart. Cnidarians, like plants, seem to have a slow rate of evolution in mitochondrial DNA and therefore need to be coded using a different segment of DNA. Most relevant to this chapter, however, is that it was very useful in identifying fish and other seafood products.

To ensure a comprehensive reference database existed for fishes, Robert Hanner, Associate Professor at the University of Guelph and Associate Director for the Canadian Barcode of Life Network, began the Fish Barcode of Life (FISH-BOL) initiative in 2005. It was one of the first taxonomically focused barcode campaigns and there has been a global effort to contribute barcodes for expert-identified reference species from around the world. At the time of writing this book, more than 10,700 species of fish have been barcoded.

Taking DNA barcoding to the market
A couple of years into building the FISH-BOL database, Hanner and his graduate student, Eugene Wong, wanted to determine whether they had gathered enough reference samples to allow them to identify unknown samples down

to the species level. So off to the markets they went. They analysed 91 samples of fish and seafood collected from commercial markets and restaurants in Canada and the US.[6] They found two things. The first was that the database was mature enough to allow them to identify the samples to species level. The second was that 23 of the sequenced samples were mislabelled in some way. Like Marko and others, Wong and Hanner found that red snapper was the most commonly mislabelled fish in the study – seven of the nine samples collected in New York were species other than *L. campechanus*. But they found that mislabelling extended well beyond this one high-value species. Atlantic halibut was being sold as Pacific halibut, Mozambique tilapia (*Oreochromis mossambicus*) was sold as albacore or white tuna (*Thunnus alalunga*), capelin roe (*Mallotus villosus*) was sold as Tobiko/flying fish roe (*Cheilopogon agoo*) and spotted goatfish (*Pseudupeneus maculatus*) was sold as red mullet (*Mullus sp.*).

Their findings stimulated new studies, all of which showed the widespread mislabelling of seafood in North America. Determined to see whether his initial tests in Canada extended beyond the Toronto region, Hanner teamed up with investigative journalists from across the country and analysed a further 236 samples from the east coast to the west; they found that 41 per cent of the samples were mislabelled.

In the US, the international organisation Oceana started conducting its own research to determine the prevalence of fraud in areas such as Boston, South Florida, New York and Los Angeles. The numbers were alarming. They worked with Hanner and his team at Guelph to conduct their analyses and between 2010 and 2012 collected more than 1,200 seafood samples from 674 retail outlets in 21 states. It was one of the largest seafood fraud investigations in the world at the time. They found that 33 per cent of the samples analysed were mislabelled. Red snapper still had the highest mislabelling rate of any species – only seven of the 120 samples labelled as red snapper were truly red snapper. Nearly 10 years after Marko's study, Oceana found little had changed; the mislabelling of red snapper is still rampant.

In 2012–13, the FDA stepped in and conducted its own study. They sampled products collected from the distribution chain across 14 states, prior to the fish arriving at the retailers (restaurants and supermarkets), targeting species that had a history of mislabelling. What the FDA found was that 15 per cent of the products they tested were mislabelled – a number quite a bit lower than the findings of Oceana, Hanner and others. Some groups, such as the National Fisheries Institute (a trade association), used these results to raise doubt about the other studies, suggesting the issue was not as widespread as they suggested. What they failed to point out, however, was that the studies targeted different areas of the supply chain. What the FDA's study did highlight was the proportion of mislabelling happening prior to the seafood getting to retailers – evidence that mislabelling is happening along every step of the supply chain.

Of course, it isn't just North Americans who are being fleeced on their fillets.[7] In Australia, studies found that 41 per cent of red emperor (*Lutjanus sebae*) and 46 per cent of dhufish (*Glaucosoma hebraicum*) were mislabelled, and 13 per cent of barramundi (*Lates calcarifer*) were also mislabelled, substituted with cheaper species such as Nile perch (*Lates niloticus*) and King threadfin (*Polydactylus macrochir*). In New Zealand, 40 per cent of 200 sampled fillets labelled as lemon sharks were other shark species, including hammerheads and bronze whalers, which it is illegal to harvest. Eighty per cent of fish samples acquired in Brazilian markets were mislabelled. Up to 36 per cent of hake sold in Spain and Greece was cheaper African species that were labelled as American and European species. In the UK, 10 per cent of 380 samples of fish collected from catering establishments was mislabelled, with swaps between cod and haddock most common. Just over 5 per cent of white fish samples collected from six major supermarket chains across the UK were mislabelled, and while this may seem like a low percentage, in terms of volume this could translate to 200 million mislabelled products sold in the UK annually. There are mislabelling examples from Ireland, Turkey, Denmark, Egypt, the Philippines, South Africa ... the list goes on. Seafood substitution is a global issue.

It's global and it's in all manner of products. Barcoding is revealing substitutions in smoked fish, dried fish, boiled fish, fried fish, fresh fish and frozen fish. What barcoding can't do, however, is trace fish back to particular regions or fish populations. Some substitutions are rather obvious – an Atlantic halibut was clearly not caught in the Pacific, for example. Finer resolution, however, such as whether an Atlantic cod was caught in the North Sea or the Baltic Sea, is beyond barcoding capability. To get to these origins, one needs to look at multiple genetic markers. This is exactly what the international EU funded project FishPopTrace set out to do. Focusing on commercial species that are susceptible to overfishing, the project began to identify minute mutations in the DNA sequence, known as Single Nucleotide Polymorphisms (SNPs), which could be used to distinguish different populations. Like the chemical fingerprints in oil that we described in Chapter 3, these are genetic signatures that become embedded in the DNA of discrete spawning populations of fish. These population markers – mutations that can be traced back to certain populations – provide a robust method for tracing fish products back to the source. But it can also be used in enforcement and as a deterrent for illegal, unreported and unregulated (IUU) fishing.

Another area where barcoding falls short is with tinned fish. The high pressure and temperature involved in the canning process clips the DNA into much smaller fragments than the 650bp CO1 sequence. The DNA is still there, but it's far more degraded. The FDA was concerned that expensive salmon species, such as sockeye, were being adulterated with cheaper salmon, such as pink and chum. They contacted Hanner to help them develop a rapid and sensitive method for testing tinned salmon. Hanner and his colleagues turned to other DNA-based methods that exploit small regions of DNA that differ between the seven species of Pacific salmon and trout, and the one Atlantic salmon species.[8] If the species is present in the sample, a chemical compound bound to the species-specific DNA segment will fluoresce. The amount of fluorescence can be measured to give an estimate of the quantity of the DNA and therefore the proportion of each fish in the sample – for example, 20 per cent pink and 80 per cent sockeye. The test is sensitive enough to detect if there's as little as 1 per cent pink salmon in the tin of sockeye, but the FDA decided to provide a little room for error and say that anything over 5 per cent adulteration would be investigated as deliberate fraud.

Farmed fish passed off as wild

The adulteration of some sockeye with pink salmon seems relatively benign in the overall scheme of things. A far more common and potentially harmful substitution among salmon is the sale of farmed Atlantic salmon labelled as wild-caught Pacific species. Once again, there is considerable economic incentive to make this substitution. Wild-caught fish sell for three to four times the price of a farmed fish. Concerns over the sustainability of farmed fin fish have created a market preference (and premium) for wild-caught animals. Investigations have found that between 15 and 75 per cent of salmon labelled as wild are actually farmed. The *New York Times* tested salmon from eight stores around the city in 2005 and six of the stores were selling farmed salmon as wild-caught. At that time wild salmon was selling for as much as US$63 per kilo (£40/kg or US$29/lb) in the Big Apple. Farmed fish was selling for between US$11 and US$26 per kilo (£7–£16/kg or US$5–US$12/lb) – a potential profit of up to

US$52 per kilo (£33/kg or US$24/lb)! Approximately 1.8 million tonnes (weighed with the head on, but the guts removed) of farmed salmon was produced in 2013. If we use the conservative value of 15 per cent mislabelling, that would suggest 276,000 tonnes of farmed salmon went out into the market labelled as wild-caught in just a single year.

And it's not just salmon. More than 100 samples collected from retailers in the UK found that 11 per cent of sea bream and 10 per cent of sea bass were farmed rather than wild as the labels claimed. These are concerning values for consumers who are trying to avoid buying farmed fish.

More alarming is that all of these farmed fish have seeped in among wild fish, which are subject to different inspection regimes. Because farmed animals are reared in pens at high densities, pesticides and antibiotics are sometimes needed to control and cure parasites and disease. As a result, farmed species undergo random testing to ensure residues of these drugs don't remain in the tissues. As with any farmed animal intended for consumption, farmed fish are subject to withdrawal periods. Before this conjures up images of anxious shaky fish circling the pens in search of a hit, this is the period of time where the fish must be drug-free before they can be killed and sold, to ensure any drugs are worked out of the system. The period of time differs greatly depending on the drug and the country.

Governments have introduced stricter regulations on many of these drugs to try and reduce their use on fish farms, yet there is some evidence that use is increasing. In 2012, a freedom of information request to the Scottish Environment Protection Agency (SEPA) revealed that the use of pesticides in the Scottish fish farming industry increased by 110 per cent over 2008 values, while salmon production had only increased by 22 per cent. It is suspected that more drugs are being used because the parasites and bacteria are becoming resistant to them. As we have learned in human health, the over-prescription of antibiotics can lead to the development of resistant strains. This has not been helped by the prophylactic use of antibiotics on many farms, particularly in developing countries. Fish are regularly dosed with antibiotics to encourage growth and prevent bacterial infections, particularly if the farm is a bit dodgy in terms of its hygiene. This creates the ideal conditions for resistant bacteria to thrive and susceptible bacteria to die off. The resistant bacteria can pass on those resistance genes and so begins perhaps one of the biggest threats we face to human health in our generation, as many of the drugs used on fish farms are also used in human health. Amoxicillin, for example, is a broad spectrum antibiotic used to treat fish for an infectious bacteria that causes, among other things, some very nasty lesions in the muscle of the animal. This drug is also one of the most commonly prescribed antibiotics for children for ailments such as ear, throat and urinary tract infections.

Farmed fish are also tested for other toxins, such as polychlorinated biphenyls (PCBs) and dioxins (by-products of some manufacturing processes), which they can accumulate from the environment but also from the fish feed. Malachite green – a fungicide – is another drug that is randomly tested for. It was once commonly used in the fish-farming industry to rid the salmon pens of parasites, but it is also a carcinogen in humans. For this reason, it was banned for use in fish farming in a number of countries, including the US, EU and Canada, by the early 2000s. Routine surveillance checks for the drug continue to be carried out and, sadly, there have been numerous accounts of the drug being found in farmed fish long after the ban.

So, between persistent organic pollutants, pesticides, antibiotics and a carcinogenic fungicide, it's not surprising that farmed fish should be subjected to more rigorous testing than wild-caught. As of 2007, after a number of incidents where imported farmed products tested positive for banned drug residues, the FDA began detaining farm-raised catfish, basa, shrimp, dace and eel coming from China; they won't release them until they're proven to be drug-free. But what if the animals aren't labelled as farmed? Farmed fish that evade testing may have unacceptable levels of toxins in them. Levels are likely to be low, but this is almost more dangerous, because eating them won't cause any acute poisoning symptoms that will go noticed, and yet there may be unknown long-term health consequences.

While misrepresenting a farmed animal as wild is dishonest (and illegal), there may be some cultural differences that make such a swap seem more justified. While many North Americans and Europeans view farmed fish as inferior, some Asian cultures prefer them. Farmed is often considered superior for sushi as there is more control over parasites, in which case swapping a farmed fish for a wild fish might actually be seen as improving the product. It doesn't justify the deception, but it does perhaps offer some insight into the minds of those committing the fraud. It's not all black and white.

'Swordfish' with a side of anal leakage

Substitutions between farmed fish and wild fish are not the only risk to human health. In 2007, Hanner and his team at Guelph were once again at work with the FDA to try and resolve a case where members of a family were hospitalised after eating home-made fish soup made with two frozen monkfish purchased at an Asian market in Chicago. Hanner's group used DNA barcoding to reveal that the monkfish weren't monkfish, but actually puffer fish (*Lagocephalus* species).[9] The FDA were surprised at the result as puffer fish is a highly regulated species in the US, but they tested the soup for the presence of a toxin associated with puffer fish, and there it was.

Puffer fish are not the fastest fish in the sea and have therefore developed some handy little defences. If they feel threatened by a predator they will take in water (or air if they are out of water) and swell in size, at the same time exposing a plethora of spikes all over their body. Most predators are deterred by the thought of trying to fit a spiky football into their mouth and give up, but those that don't are rewarded by a rather unpalatable mouthful. Puffer fish contain tetrodotoxin – a neurotoxin – in their tissues, particularly in the liver, ovaries, intestines and skin. The toxin is thought to be produced by bacteria living in the gut as well as on the skin of the fish. While the toxin isn't usually enough to kill the predator, it may deter them from ever eating puffer fish again. For some predators, the toxin is lethal, while others, such as tiger sharks, seem unaffected and happily munch on puffer fish. However, as little as two milligrams of the toxin is potentially enough to kill a human (though this varies depending on the person's age, weight, health and sensitivity). Hence the strict regulations. Only certain parts of certain species are permitted into the US through the JFK International Airport in New York. These fish have to be prepared by authorised facilities, certified for safe consumption and sold only to restaurants belonging to the Torafugu Buyers Association.

Yet, here were two puffer fish that had escaped all of this under the guise of monkfish. And to make matters worse, there were unusually high levels of the toxin in the muscle of the fish, which isn't usually the case. Luckily, medical staff and numerous agencies responded quickly and managed not only to treat the family appropriately but also to track down the source of the fish. The supplier was placed on the FDA's Import Alert list for misbranding. One of the family members, who had consumed three milligrams of the toxin (a potentially lethal dose), suffered from chest pains, vomiting, numbness and profound weakness in her lower extremities. She required three weeks of inpatient therapy before being discharged to a long-term care facility for further rehabilitation. In summary, if you experience numbness or tingly lips after eating so-called monkfish (or any other fish for that matter), seek medical advice immediately.

Seafood substitution also presents a health risk to people with seafood allergies. People rarely have allergies to all seafood, just particular groups of seafood. Someone with an allergy to fin fish, for example, would no doubt be very unhappy to buy a crab product, only to find out that it is actually a surimi-based product made with hake. Even among fin fish, people may have an allergy to salmon but not tuna. It turns purchasing fish into a rather risky game of allergen roulette.

Somewhat less lethal but nonetheless unpleasant is the substitution of oilfish, snake mackerel or escolar for white tuna, swordfish and even Atlantic cod. The oilfishes contain high levels of wax esters, compounds made up of a fatty acid and a fatty alcohol that are largely indigestible. So, after a lovely romantic fish dinner, you may be in for a night of keriorrhea. Put bluntly, this is uncontrollable anal leakage of an orange to greenish oil – maybe not what you had in mind for those fancy new third-date underpants. Add a little vomiting and abdominal cramps to this and it makes for a lovely evening all around. In 2007, Hong Kong consumers who were excited to be paying four to five times less for 'Canadian Atlantic cod' had first-hand experience of the effects of oilfish, with over 600 people falling ill after eating the fish. Japan and Italy have banned the sale of escolar, while countries like Australia, Canada, the UK and the US have opted for more of a public education/fact sheet approach to protecting consumers, while still permitting its sale. Might be worth familiarising yourself with oilfish and escolar!

Fish laundering is bad for the environment
Seafood fraud is hurting our wallets, our health and sadly also our environment. Just as shell companies can help to

clean drug money, species substitution helps to launder
the products of IUU (illegal, unreported and unregulated)
fishing. These practices undermine conservation measures
put in place to help conserve species by catching more fish
than is allowed, fishing in closed areas or out of season,
using gear that is prohibited, or catching outside of the
allowable size limits. In 2004, a US company, Neptune
Fisheries Inc., was convicted of conspiring to import
86,182kg (190,000lb) of undersized frozen spiny lobster
worth more than two million US dollars (over £1.2
million). Two years later, a husband-and-wife team
operating Anchor Seafood Inc. were convicted for
smuggling 7,500kg (16,500lb) of undersized spiny lobster
into the US from Jamaica. They organised 40 illegal
shipments between January 2000 and January 2001, valued
at approximately US$229,000 (£144,630), violating
harvest restrictions in both Jamaica and Florida. Both
cases were uncovered by the US's National Oceanic
and Atmospheric Administration (NOAA) Fisheries
Office for Law Enforcement. Special Agent Scott Doyle,
an investigator for NOAA for over 27 years, said in an
interview to the *Baltimore Sun*, 'If I was going to be a
criminal, I would be in the fish and wildlife smuggling
business. Nobody has any idea what's going on. They just
buy fish.'[10]

IUU is estimated to be worth €10 to 20 billion (£7.1–
14.2 billion, US$11.4–22.8 billion) per year and is equal to
at least 15 per cent of the reported global catch. In some
parts of the ocean it is thought that as much as 37 per cent
of the catch is illegal and for some high-value species, such
as tuna and swordfish in the Mediterranean and sharks
across Europe, IUU fishing may be as high as 50 to
75 per cent. It is a major contributor to the depletion of fish
stocks, and connections have been made with drug
trafficking, human trafficking and other illegal activities.
It is the biggest global threat to the sustainable management
of our fisheries. The EU has banned imports from fishing

vessels registered to countries that aren't taking sufficient action against IUU, including Belize, Cambodia and Guinea. Curaçao, Fiji, Ghana, Korea, Panama, Togo and Vanuatu all received formal warnings of a potential ban in 2012. Since then, Belize has had its ban lifted owing to the steps the country has taken to help address illegal fishing. Fiji, Panama, Togo and Vanuatu have also been de-listed by the EU for their positive actions.

While seafood fraud is helping to launder dirty fish, it's also undermining any efforts consumers are taking to make sustainable seafood choices. Many groups around the world, such as the Marine Conservation Society (UK), the Australian Marine Conservation Society, World Wide Fund for Nature (WWF), the Monterey Bay Aquarium (US) and the David Suzuki Foundation (Canada), have developed pocket guides to help consumers make better seafood choices to support sustainable fisheries. Yet, these efforts are undermined if the species or the location where it was caught is falsely labelled. For example, Atlantic halibut, *Hippoglossus hippoglossus*, was heavily overfished in the nineteenth and twentieth centuries and has never recovered. It is listed as Endangered on the International Union for Conservation of Nature (IUCN) Red List. Today there is no directed fishery for this species within US federal waters with the exception of some small-scale harvests off the coast of Maine. The Monterey Bay Aquarium's Seafood Watch guide recommends avoiding Atlantic halibut, yet the species is caught as by-catch in US and Canadian ground fisheries. As there is little market for these endangered species, they are often labelled as the closely related Pacific halibut, *Hippoglossus stenolepis*. Pacific halibut fisheries have been better managed; a number of them have been independently certified by the Marine Stewardship Council (MSC) and are listed as good alternatives or better choices in seafood guides. To confuse matters, the Canadian Atlantic halibut fishery achieved MSC certification in 2013, which introduces the potential for non-sustainable US-caught populations to be passed off as an MSC-certified fishery.

It is incredibly disempowering to think that the efforts we, as consumers, go to in order to inform ourselves and feel able to interrogate our fishmongers about the fish we're buying are futile at least once in every three purchases. Would we be so tolerant of this if mountain gorillas or Ganges River dolphins were being ground up and making their way illegally onto our plates?

Fish out of water

Of course seafood fraud isn't only about species substitution. There are many other ways to cheat consumers. Perhaps the most common form of seafood fraud, though less talked about, is short-weighting. This is when processors increase the weight of the product by adding a little more breadcrumbs or batter to the fish fingers or more ice to the prawns and include this in the net weight of the product. In 2012, the *Boston Globe* surveyed 43 samples of seafood collected from supermarkets across the state of Massachusetts. They found that 20 per cent of the samples glazed in ice weighed less than the packaging claimed – a practice known as over-glazing. Another study of 240 shrimp samples imported into Europe from South-east Asia found that half of them were short-weighted, with some samples weighing as much as 28 per cent less than the weight declared on the label.

Another method of short-weighting is to over-soak the seafood products in sodium tripolyphosphate (STPP). This is a preservative commonly used in the food industry (labelled as E451 in Europe) and it helps keep seafood firmer and smoother by helping retain moisture in the flesh of the fish. STPP and other similar additives are a legitimate way of keeping seafood from drying out in the freezer. However, over-soaking in STPP causes the flesh to absorb and retain an unnatural level of water, which the customer ends up paying for. Scallops are particularly prone to this practice as it bloats them up into tempting plump portions. As a result, regulators have put in place specific regulations on the water content of scallops. Canada doesn't permit the use of phosphate additives in scallops at all and considers any scallops with a moisture content above 81 per cent to be adulterated with water; the average water content of a scallop harvested fresh from the ocean is about 75 per cent. The US and EU require that phosphate additives and any added water be declared on the label. This includes any water inadvertently added during storage by the uptake of melted ice water. Leaving scallops sitting on ice for 10 days in a commercial storage facility can increase moisture content by 2 per cent.

Added water in our seafood may not pose any health risks and the cost to the individual consumer can be relatively low. It may be a matter of a few pennies here and there, but one US investigation in 2010 across 17 states found consumers paying as much as US$50 per kilo (£31.66/kg or US$23/lb) for unwanted ice. Multiply this through the industry as a whole and these costs are extraordinary – millions of US dollars each year. While this clearly constitutes economic fraud, the priority for seafood inspections will continue to be around food safety. Also, because glazing and soaking are legitimate processing techniques, it can be difficult to detect when the line has been crossed from providing customers with a better product to providing them with a whole lot of water.

As consumers it's virtually impossible for us to determine whether we're paying for excess water at the point of purchase. This water doesn't get released until cooking, when you watch in horror as your sautéed fish dish shrivels and falls apart before your eyes as your dinner guests sip their wine. A fish fillet over-soaked in STPP will shed the excess water during cooking and release a milky white liquid in the pan. This liquid can wreak havoc on any sauce you might be using, and when the liquid is released, the fillet becomes very fragile, often breaking apart despite gentle handling. Over-soaked scallops will behave similarly and may not sear properly because of the added water. These revelations (if we even have them) may frustrate us, but will it be enough to motivate us to complain to the retailer or to the processor, or to report it to the responsible legal authorities? Let's face it, most of us will probably have forgotten about it by the time dessert rolls around. You're unlikely to get too emotional over a few pence-worth of water added to your fish fillet, just as you are unlikely to cross the street to pick up a couple of wayward pennies on the side of the road. But would you be willing to cross the road with a thousand people to get your share of a couple of million pounds? In 2010, over 29 per cent of fish for human consumption was sold as frozen product – worth just shy of US$55 billion (£35 billion). Those pennies add up.

Closing the net on seafood fraud

There's no denying that seafood fraud is rampant, so the next question is what is being done about it. It would seem that some governments are starting to wake up to the issue of seafood fraud and its risks to consumers, the extent of economic fraud and its role in supporting IUU fishing. As of December 2014, EU rules changed so that it became mandatory to provide the scientific name as well as the commercial name for species on all unprocessed and some processed fishery products (mainly smoked and salted products). Canned, composite and breaded products do not need to provide a species name, but they do have to provide a quantity on the ingredients label. A tin of mackerel, for example, must have 'mackerel (75%)' in the ingredients list. In July 2014, President Obama established a national task force to combat seafood fraud and IUU fishing. The recommendations of the task force, released in December of the same year, included better systems for gathering, sharing and analysing information about the seafood products entering the US as well as an effective seafood traceability programme.

More random unannounced testing of seafood products along every link of the production chain would increase the risk for fraudsters, and stiff penalties when caught would probably help tip the scales on seafood fraud. DNA barcoding has provided a global standard for identifying mislabelled species, but if there are no resources available to do the testing, it's useless in battling food fraud. When it comes to inspections, most countries are just trying to manage their risk. How they choose to do that may vary, but it really is about targeting items that present the highest risk to the public. Every country has particular products, importers and countries that are flagged due to a previous incident or known conditions in that country. These shipments will always be inspected. For example, the CFIA in Canada inspects 100 per cent of fresh or frozen tuna (all species) imported from India by Moon Fishery India Pvt. Limited – but they're only looking for *Salmonella*. Less attention is generally given to imports from countries where there are existing agreements. Fishery products move relatively unrestricted between EU countries, for example. Some countries, such as Canada and New Zealand, try to maintain standards by specifying that seafood importers be licensed, while others, such as the US, look for certification in the foreign exporters. Australia inspects 100 per cent of risky foods and then reduces this rate to 5 per cent once there is a demonstrated history of compliance. The FDA in the US inspects 1 to 2 per cent of fish imports and less than 1 per cent of these will have any analyses done. An inspection can entail anything from going over paperwork, to climbing into the shipment (quite literally) and pulling out samples to make sure they match the paperwork, to opening the product up to conduct testing.

In fact, it is very difficult to gauge just how much testing is being conducted in a lot of countries. The FDA inspection rate seems low, but US Customs and Border Protection and NOAA are also doing inspections. The industry itself is conducting internal inspections; academics and independent watchdog organisations, journalists and conservation groups are doing testing. At the time of writing this book, US Senator David Vitter had introduced the Imported Seafood Safety Standards Act. If the bill is passed, it will mean more inspections and testing of imported seafood and increased penalties for mislabelling. Bills such as this can make authenticity testing of equal importance to pathogen testing.

Governments and stakeholder consortiums are also taking steps to standardise and harmonise testing around the world to better cope with the global movement of seafood. In 2011, the FDA released its own DNA barcode database. The FDA is part of the Barcode of Life consortium and has adopted its methods to target the CO1 gene sequence, but they wanted to develop an extremely rigorous database so that any evidence gathered could stand up in court. They worked with the Smithsonian Institution's Laboratories for Analytical Biology and the Division of Fishes to ensure the database of 250 commercially important species were all identified by experts from the Smithsonian. While the FDA's database may be more rigorous in terms of identification than BOLD, it isn't as comprehensive yet, which may be limiting. More than 20 FDA analysts are now trained to do DNA barcoding. Europe is also considering adopting the Barcode of Life protocols. Labelfish is a multi-national project that was set up to develop a standardised approach to testing seafood across Europe. They are recommending that barcoding has consistent and straightforward protocols that can be used by most authenticity labs across the EU, and as countries such as Canada and the US are using these methods, it provides a common international approach to combating fraud.

With governments limited by resources and millions of tonnes of fish moving about the world, some groups within the industry are taking matters into their own hands. Fishermen are seeking ways to make their product traceable. In the US, fishermen in the Gulf of Mexico have developed the Gulf Wild brand of fish, which are responsibly caught and safety tested. In the interests of fraud prevention, Gulf Wild fish are tagged with a unique number that enables consumers to go online and use that number to find out who caught the fish and where.

In 2009, fishermen on the west coast of Canada approached the non-profit organisation Ecotrust Canada about setting up a system of traceability for their product. Ecotrust Canada launched ThisFish in October 2010. It's an extensive database that, like Gulf Wild, uses a unique code associated with each fishery product to allow consumers to learn more about the product – who caught it and where, as well as what equipment they used. The code might be found on the label of a can of salmon or on a tag around a lobster claw. The programme is voluntary and fishing vessels simply have to register. Processors and distributors can also register and upload information to the system so that consumers can get information regarding every link along the seafood supply chain. There's even an option for consumers to leave feedback for the harvesters and processors. It's a sophisticated database and a simple idea that helps reconnect fishermen with consumers and share the story behind each seafood product. And the stories behind these Canadian products have an international reach. Shortly after ThisFish launched, they were seeing Canadian lobsters being tracked by consumers in Belgium, the Netherlands, Norway, France and Spain. ThisFish has expanded beyond Canada, working with fisheries in Indonesia, Iceland, the Netherlands and the US.

As well as improving traceability, the industry is conducting authenticity testing as part of its protocols. A seafood importer based in Victoria, Canada started incorporating DNA testing into its business practices in 2011. Tradex Foods Inc. started DNA testing seafood that it was bringing in from China for its Sinbad house label. The samples are taken by its China office and flown to a US-based testing facility while the fish itself is making its way across the Pacific by ship. It costs the business between CA\$700 and CA\$2,100 (between £360 and £1,078) each month to do the testing, but it's part of their business model not only to have a rigorous quality control process, but also to help eliminate fraud.

Tradex uses a private laboratory in Illinois called ACGT, Inc., which is making a niche for itself with seafood importers, wholesalers and retailers who view DNA testing as an investment in their businesses. As consumers become more infuriated by fraud and begin to look for proof of authenticity or traceability in their products, this kind of initiative could help give forward-thinking businesses a market advantage.

Just as we complete this chapter, researchers at the University of South Florida have come up with a hand-held device known as Grouper Check, which uses DNA-based methods to check whether your grouper is really a grouper. The instrument sells for about US\$2,000 (£1,268). We are likely to see more of these technologies hitting the market over the next few years, which could make it more cost effective for industry to conduct more internal testing.

As consumers, there are steps we can take as well to help reduce our exposure to seafood fraud, whether it's species substitution or short-weighting. First and foremost, know your product. If you are out to buy monkfish, search for some pictures of the species before you go and, most importantly, buy fish that have some identifiable features, such as skin and maybe even a tail fin. If you can, buy fresh to be sure you aren't paying for any additional water. Second, know your seller (and any others who have touched your fish along the way). If you are lucky enough to live in a place where you are able to buy direct from the people catching the fish, then do so. Go back to those you like and that provide a good product for a fair price because building a relationship with the people who sell you food will probably be more valuable in understanding your food than any label could ever be. If you can't buy directly, then look for tracing organisations where you can check provenance, such as ThisFish or Marine Stewardship Council. Third, know your season. Just like fruit and vegetables, fish have seasons too. The substitution of farmed salmon for wild is far more common when wild salmon aren't in season (approximately October to April).

There is no doubt that things become more complicated when dining out. The Oceana study found that 74 per cent of sushi bars, 38 per cent of restaurants and 18 per cent of grocery stores they took samples from were selling mislabelled fish. Technologies such as Grouper Check may be a prelude to consumer-targeted kits; a dipstick test much like a pregnancy test is possible, for example. But instead of peeing on it you hold it to your swordfish steak: two blue lines confirms you've been served swordfish, while one blue line means it's something else. Of course, just because such a test is possible doesn't mean there would ever be a market for such a thing. Ten quid for a one-use test that tells you your fish is what the menu says isn't likely to be a big seller, and what would it say about the state of our food industry if it was?

Despite the rampant mislabelling in sushi bars and restaurants, there are establishments that are reputable and care not only about providing quality food to their customers, but also about making a commitment to the health of the ocean on which they depend. Moshi Moshi in the UK comes to mind immediately, as founder

Caroline Bennett has been at the forefront of campaigns to protect fish stocks. She works alongside ethically minded local fishermen to provide seasonal quality products to her London sushi restaurants. She knows her seafood. And if you care about the environment, your body and your wallet, this is the best advice: get to know your seafood.

CHAPTER FIVE

What's Your Beef?

Before there was horse meat in our burgers, there was Maggot Pete, perhaps one of the most notorious meat fraudsters in recent UK history. Peter Roberts was a businessman whose first enterprise was a maggot farm, where he earned his unfortunate nickname. In search of bigger profits, Roberts opened a poultry slaughterhouse, Denby Poultry Products, located in Derbyshire, England. He began using the waste products from his slaughterhouse legally in processed pet food to increase the profitability of his business. But he recognised an even more lucrative business opportunity: redirecting waste products back into the human food market. As well as collecting low-risk waste – animals not fit for human consumption, mostly for aesthetic reasons – Roberts began purchasing diseased and contaminated poultry from slaughterhouses. Some of these animals had died of unknown causes and could be carrying transmissible diseases. The slaughterhouses had been paying about £80 (US$126) per tonne to have these high-risk waste products taken away and destroyed, so when Roberts started offering £25 (US$39) per tonne to relieve them of the waste, it was an easy decision. Workers in Roberts's factory would then trim away the undesirable parts, wash the chicken with bleach to remove slimy layers and discolouring, and finally package it up for sale to hospitals, schools, restaurants and leading supermarkets.

Between 1995 and 2001, Roberts and his team turned just under half a million kilos (over one million pounds) of condemned poultry out into the human food supply and built up a client base of around 600 customers. The operation earned the ringleaders a combined estimated total of £1 million (US$1.5 million) over the six years.

In 2000, an anonymous tip led environmental health officers in Derbyshire to begin investigating Denby Poultry Products. By 2001, they had enough evidence to raid the premises, where they found skips filled with green decaying poultry and a large pool of raw sewage in the middle of the processing plant. The investigation lasted over two years and involved more than 100 police officers and 50 local authority environmental health officers. The investigators unravelled the threads that connected Denby Poultry Products to over 1,000 other food manufacturers, wholesalers and retailers around the country. They found fraudulent European health stamps, used to trace animal origin food products in the EU, which had been used to stamp the condemned chicken. Sainsbury's, Tesco, Kwik Save and others began to recall potentially contaminated products – a process that cost the industry about £1 million (US$1.5m).

Six people were convicted in 2003. In addition to Roberts, two former managers and an occasional worker at Denby Poultry were given jail sentences, as well as two people that worked at MK Poultry, which is a food processor in Northampton that supplied the meat to retailers and added the European health stamp. Roberts was sentenced to six years in prison, but fled the country before the trial was over.

In 2007, Maggot Pete was found soaking up the sun in the Turkish Republic of Northern Cyprus, which conveniently has no extradition treaty with the UK. Perhaps the £435,183 (US$686,923) he had personally made on the scam hadn't quite been enough to sustain the family lifestyle, or perhaps his wife, Shari Roberts, just needed to get out of the house, but whatever the reason, Mrs Roberts was found working in the local estate agent. It was ultimately through her that Roberts was found. Despite the lack of an extradition treaty, Roberts arrived at Stansted airport within a month of being found, and shortly thereafter he appeared at Derby Crown Court, where he received his sentence. As well as his six-year term, Roberts was ordered to pay back more than £167,000 (US$263,600), which was the estimated amount left from his profits.

Roberts certainly wasn't the first to see this opportunity for repurposing meat. Other operations even bigger than Maggot Pete's have come before and since, but whether it was due to his catchy nickname or his flight to Cyprus, Roberts' scam became one of the more notorious. In 1995, about the same time Roberts was starting his operation, Rotherham Council began investigating a secret operation at Wells By-Products factory in Darlton, Nottinghamshire. The Darlton gang were using the same business plan: to turn condemned poultry destined for pet food back out into the human food supply. In a three-year period, they sold more than one million kilos (over two million pounds) of rotten chicken and turkey to retailers and distributors across the UK. Those involved made about £2 million in profit (over US$3 million), while the investigation cost Rotherham Council half a million.

Fresh meat is highly regulated, but there was a loophole at the time that made these crimes possible: poultry was not included in the regulations that require condemned meat to be stained with an indelible dye to prevent it from finding its way back into the food chain. In 2001, following these scandals, the FSA started a public consultation to tighten up the regulations, including a proposal to stain high-risk and low-risk poultry by-products. By-products are considered to be high risk if there's a chance they're carrying transmissible diseases (such as BSE) or contain residues of prohibited substances (such as pesticides) or environmental contaminants (such as PCBs). These are classified as Category 1 animal by-products and are incinerated or sent to landfill after heat treatment. Category 2 by-products are also high risk and include animals rejected from abattoirs owing to their having infectious disease, or animals containing residues from legal treatments (such as antibiotics). These materials can be recycled for non-feed purposes such as oleochemical products (chemicals derived from animal fats). Low-risk by-products include those that are past being fit for human consumption or have been withdrawn from the market for reasons other than food safety concerns. These can get recycled into animal feed by licensed Category 3 processors. While many applauded the proposal to mark these by-products, poultry processors protested that these changes targeted legitimate British poultry processors that already had rigorous hygiene enforcement. Compulsory staining would increase their costs by as much as £30–£60 (US$47–$95) per tonne, depending on how much labour was required to prepare the by-product for staining. A carcass, for example, would require many deep incisions to ensure the stain penetrated the muscle. The cost of staining high-risk by-products for a typical large processing plant that's moving over 41 million birds through its premises annually would be about £46,560 (US$73,372) a year. This cost would increase to £832,980 (US$1,312,737) per year if both high-risk and low-risk waste staining requirements were in place.[1] The industry was concerned that these additional costs would lead to higher-priced British poultry being outcompeted by cheaper imports.

In 2002, regulations were introduced that require high-risk poultry by-products to be stained with a blue dye. Lewis Coates, the environmental health officer who led the team investigating the dubious processing in Darlton, suggested in an interview with the BBC that these regulations are still insufficient. What happened in Rotherham would still have happened under the new regulations: 'It wouldn't have made a difference; the meat would still have come through.'[2] Low-risk products, which are used in pet food, aren't covered by the legislation and therefore don't require staining. The separation of the waste into various categories of risk is up to the processors, and while inspectors are responsible for checking the waste as well, there is little time to do so.

Luckily, neither of these poultry scams resulted in anyone becoming ill – despite some of this meat being sold on to institutions caring for some of the most vulnerable people in society, namely hospitals and schools. One can't help but think that had a few people become ill as a direct result of these scams, the offenders might have received tougher penalties.

A rotten history

If we go back to the mid-nineteenth century, the sale of putrid meat was a weekly event. In Britain, shops weren't allowed to sell anything on a Sunday and all shops had to close by midnight on Saturday. Therefore, anything that wasn't going to keep until Monday was sold off at a discounted price. Saturday afternoon was also when many workmen received their weekly wages. The marrying of cheap food and money in the pocket resulted in the Saturday-night shopping phenomenon, where the working class would do their buying between 10 p.m. and midnight, when their money went a lot further. Shopping by candlelight also meant that the merchants could get away with things they couldn't have during the light of day. A layer of fresh fat would be added to rotting meat to make it appear fresh. The meat the working class was buying on Saturday night was already past its 'use by' date before they bought it, let alone got it home and cooked it for the Sunday roast. But having paid for it, people felt obliged to cook it anyway, hoping for the best. There were meat inspectors on the hunt for tainted meat in these Saturday-night markets – in fact, it was one of the few types of food being policed at the time. The inspectors would hand out fines and confiscate tainted products. They would take diseased animals from the markets and hand them to the police for destruction – sometimes quite publicly. Yet, despite their efforts, the sale of rancid meat continued. Poor people were looking for cheap meat; they knew the deals were too good to be true, but it was all they could afford, and sellers were looking to offload products that were past their prime. Criminals and victims were in some sort of odd unspoken collaboration.

On the other side of the Atlantic, the release of Upton Sinclair's 1906 novel *The Jungle* awakened Americans to the state of their meat processing. Sinclair's intention had been to draw the nation's attention to the exploitation of US immigrants through the eyes of his main character, Jurgis, a Lithuanian man living and working in the stockyards of Chicago. However, it was the grotesque conditions in which meat was being processed that truly revolted the nation. Sinclair's evocative descriptions of mouldy white sausages, piles of meat covered in rat droppings, and even the rats themselves being swept into the hoppers where sausages were being churned out for home consumption were disturbing. The imagery is enough to convert even the most dedicated carnivore into a vegetarian. President Theodore Roosevelt, understanding very well the power of the meat and packing industries, sent two commissioners, Charles P. Neill and James Bronson Reynolds, to investigate whether Sinclair's descriptions were purely an author's creative licence. The commissioners confirmed all Sinclair's claims, except for one: apparently it was an exaggeration that workmen were falling into the vats and being rendered into pure leaf lard (the highest grade of lard). Sinclair's novel and the subsequent report from Neill and Reynolds were in part responsible for the Federal Meat Inspection Act of 1906, which would help prevent adulterated or misbranded meat and meat products from being sold as food, and would try to raise the sanitary conditions of slaughterhouses and meat processing plants. Sinclair later stated, 'I aimed at the public's heart, and by accident I hit it in the stomach.'

History suggests that people like Maggot Pete and his conspirators are simply carrying on an age-old tradition of repurposing old food. Condemned meat continually makes its way back into the human food chain and no country seems to be immune. In 2006, a 74-year-old German meat distributor committed suicide when he was linked to laundering expired meat back into the food chain. Police impounded more than 150 tonnes of expired meat – some more than four years out of date – from the company. In the summer of 2014, the fast food chain McDonald's found itself embroiled in a tainted meat scandal. Shanghai Husi Food Co. was inspected after an undercover journalist exposed unhygienic handling of meat. The investigation revealed expired beef and chicken products being processed and repackaged with new expiration dates. The company supplies chicken and beef products to McDonald's, Papa John's, Burger King, Starbucks, KFC and Pizza Hut in China as well as about 20 per cent of McDonald's products in Japan. The Japanese branches of the fast food giant stopped importing their meat from China and took the opportunity to introduce a tofu and fish version of the McNugget.

Is it possible to view these scams as an attempt to reduce waste? We've all done it on a small scale, whether our motives were to avoid wasting food or to avoid wasting money. We've pulled out that piece of meat from the fridge, looked at the 'use by' date and sighed. Then, we have cautiously opened up the package and had a sniff. Barring no ill odours, we may poke and prod the meat with our fingers, maybe even lift it out of the packaging and take a good 360-degree view. After another glance at the label and some mental maths to count back the days, we make a decision either way – bin it or eat it. If it's the latter, we probably cook the hell out of it, killing all signs of potential life and any flavour along with it. Then, we spend the next 24 hours in a heightened awareness of all our bodily functions.

It's hard to say how many people get sick each year from eating expired meat as publicly available statistics do not separate illness as a result of eating expired meat from something like bad hygiene or insufficient cooking.

The FDA in the US and the FSA in the UK are more concerned about tracking the prevalence of infectious agents (bacterial, viral, chemical and parasitic), whether it's *Salmonella* or *Listeria*, and identifying new and particularly virulent strains of bacteria such as *E. coli* O157:H7. Whether the bacteria is there because the food wasn't cooked correctly, or because of cross-contamination between cooked food and uncooked food, or because of an employee's poor washroom habits, is of slightly less concern than what kind of bacteria it is, how widespread it is and how the illness can be contained. It boils down to a matter of risk management. US outbreak statistics from 1998 to 2008 show that 22 per cent of foodborne illnesses were attributed to meat and poultry. Just as a comparison, vegetables were responsible for 34.2 per cent and shellfish a mere 3.4 per cent.[3] Some of that 22 per cent may be the result of eating expired meat, but the vast majority is not.

While trying to find examples of people becoming ill as a result of expired meat scams, we came across pages upon pages – entire forums even – devoted to the discussion of eating expired meat. Some tout the health benefits of eating purposely rotted meat. The late Dr Aajonus Vonderplanitz, for example, who turned to a diet of raw rotting meat after being introduced to it by some of the indigenous peoples of Alaska. He called it the primal diet. Some discussions are prompted by panicked mothers who have just inadvertently fed their families minced beef that expired a week ago. Others are queries about how to prepare expired meat and how long, exactly, can it be expired before things get really dangerous. It became clear to us while writing this chapter that there are a large number of people out there intentionally buying reduced meat that's close to expiry and genuinely researching ways to reduce their exposure to food poisoning. It is the modern-day Saturday-night market.

The FSA estimates that at least 40 per cent of consumers are prepared to eat food that is past its use-by date. And as with all aspects of life, some people are greater risk takers than others. Whether motivated by money, following a primal diet or a desire not to waste the steaks rediscovered in the back of the fridge, people are making the decision to eat expired meat daily. The difference, of course, is that *they* are making that decision. They are using a number of different sensory clues (the smell and look of the meat) and their own personal past experiences (some people have stomachs of steel) to make their decision. It becomes fraudulent when this decision is taken away from the consumer, when consumers are sold something that isn't what it claims to be, or when the tell-tale signs of bad meat have been cut away and masked by chemicals. It is then that consumers should be armed with the knowledge that if a deal seems too good to be true then it probably is – particularly in the world of meat.

Tight reins
Meat is a targeted commodity for fraud because, quite simply, it's a high-end product. And, as well as being expensive, there is an increasing appetite for it. Our ever-increasing population on this planet, and in particular an expanding middle class (in terms of both numbers and girth), is creating an unprecedented demand for animal-based protein. In 1964, the global average for meat consumption was 24.2kg (53.3lb) per person per year. Thirty years later, this average had increased by more than 10 kilos (22lb) – equal to about five chickens more a year. Meat-loving countries such as Australia, US, Luxembourg, New Zealand and Argentina bring the global average up with annual per capita consumptions of over 100kg (220lb). But it is a shift in countries where people have traditionally eaten very little meat that is creating the biggest change. The average consumption of meat in China in 1961 was 3.6kg (7.9lb) per person per year, but by 2011 this average had increased to over 50kg (110lb). With a population of over 1.35 billion, this is bound to have an impact on global demand for meat.

Though there are significant profit margins to be had in shady meat, there is also considerable risk. Meat is one of the most highly regulated food products. There are numerous reasons for this: there are welfare concerns for the animals; carcasses carry disease that can be transferred to consumers; the nature of the slaughtering process has the potential to introduce bacteria that can be harmful if eaten; and, as we've mentioned, it's vulnerable to fraud.

In the UK, food hygiene legislation requires that all meat processing plants be approved by the FSA or, in some cases, the local authority. Slaughterhouses, cutting plants and establishments that handle game are subject to routine inspections and audits as well as veterinary controls. Every certified abattoir has an official government-employed veterinary surgeon and government meat inspector present in the plant and on the processing line any time the plant is in operation. When the animals arrive at an abattoir they are given an ante-mortem health inspection by the veterinarian. Their paperwork is also inspected. The animals are stunned and then slaughtered. Once the animals have been skinned (or not, in the case of poultry) and the organs removed, the veterinarian conducts a post-mortem inspection to detect diseases that may be a risk to public or animal health, to look for residues or contaminants that exceed legislated levels, to assess the risk of non-visible contamination and to look for anything else that may lead to the meat being declared unfit for human consumption. They also look for evidence of injuries that may indicate animal welfare concerns. Nearly 940 million animals are killed for human consumption in the UK in a single year, and over 850 million of these are chickens. This is a lot of inspections. Many of these animals will be rejected for human consumption for one of three reasons: the animal contains pathogens harmful to humans, it hasn't met certain legal requirements, or it is aesthetically unpleasing. Most of the meat rejected in the UK is discarded for aesthetic reasons. Between 2012 and 2014, over five million red meat animals (this doesn't include poultry, but does include pigs) were recorded as having conditions such as pneumonia, abscesses, septicaemia (blood poisoning), tumours and tuberculosis. Many of the diseases identified present no risk to humans, yet they may change the look of the meat in a way that would make it unappealing to consumers. For example, *Cysticercus ovis* is the larval stage of the tapeworm parasite *Taenia ovis*, which finds comfort in the intestines of dogs and wild carnivores, using sheep and goats as intermediate hosts. It is not a threat to humans, yet over 190,000 animals were rejected for *C. ovis* cysts between 2012 and 2014. People are more likely to pick up a nasty parasite from letting their dog lick them on the mouth ... but we digress. The point is that these rejection rates are a reflection of tight regulations.

Horse on the loose

Yet, despite tight reins, the horses got loose – as anyone who ate a beefburger or lasagne from some of the leading UK supermarket chains in early 2013 probably knows. Irish authorities announced that routine testing using DNA-based methods had detected horse meat in beef products in December 2012, but they couldn't state the quantity. Further tests using quantitative PCR analysis estimated that frozen Tesco value-brand beefburgers contained up to 29 per cent horse meat.

Quantitative or real-time PCR amplifies targeted sections of DNA and is sensitive enough to detect adulteration levels of 0.1 per cent. Most importantly, it can quantify the different species detected. DNA is extracted from a sample of burger labelled as 'all beef'. The sample DNA is then mixed with all the ingredients that are necessary to help make copies of the targeted region of DNA. This includes: species-specific primers that bind to the area to be replicated; enzymes, which are the agents that facilitate the copying process; and free nucleic acids, which are the building blocks used to make the copies.

A fragment of DNA, known as a probe, is also added to the mix. This probe is labelled with a fluorescing molecule and has been designed to bind to the copied DNA. The whole mixture undergoes a sequence of heating and cooling cycles, which control the activity of the enzyme that facilitates the copying. If, for example, the horse primer recognises horse DNA in the burger sample, it will bind to that site and the DNA will be replicated. If horse isn't present, the primer won't react. With real-time PCR, the DNA copies can be measured as the reaction happens by measuring the fluorescence of the probes that have attached to the replicated DNA. Then working backwards, the measure of fluorescence and the number of replication cycles can be used to estimate the quantity of DNA in the original sample.

As always, however, there are limitations. First, the most processed meats are also those most likely to have undergone treatments that have the potential to degrade the DNA. This obviously affects the ability to replicate it. Second, the amount of extractable DNA can also differ between tissues. This means that if horse organs and cartilage have been added to the burger rather than just muscle meat, calculating the relative quantity of DNA may be flawed if using a calibration standard of just muscle meat. Third, the test can only detect DNA fragments recognised by the primers added. The analysts were specifically trying to quantify horse in the beefburgers, so they would have added primers specific to horse DNA. There could actually be other species in there that wouldn't be detected unless primers for them were also included. In other words, you're not going to find a rat (sometimes literally) unless you're looking for it.

Horse was also found in Findus lasagne and Tesco brand spaghetti bolognese. The FSA asked supermarkets and branded manufacturers to send all processed beef products for testing and a further 2,501 DNA-based tests were conducted to look for adulteration of beef products. Seven products were found to be affected in the end; frozen lasagne and spaghetti bolognese manufactured by Comigel was up to 100 per cent horse meat. There was undeclared pork found in the meat as well – 85 per cent of the tested products had undeclared pig DNA – but that didn't make nearly as good a headline. Europeans wanted to know who took the beef out of their burgers and bolognese. As investigators tried to trace the source of horse through the supply chain, it quickly became evident that it was the complexity of the chain itself that was part of the problem.

For those who may not be well versed in Horsegate, this is a very brief overview of one strand of this complex chain unravelled by investigative journalists at the *Guardian*. Horses, some of which were sick and most of which were badly treated, were being illegally smuggled from Ireland to Scotland and on into England. Some of these horses ended up at the UK's Red Lion abattoir, which had previously earned a dubious reputation for illegal and inhumane practices. Horse meat from the Red Lion was purchased by Dutch meat wholesaler Willy Selten BV. Investigations of the wholesaler led to the arrest of owner Willy Selten in May 2013. Selten's business had received more than 300 tonnes of horse meat between 2011 and 2012 from England, Ireland and the Netherlands, yet the books only showed records of beef. The company was ordered to recall 50,000 tonnes of meat that it had sold to 16 European countries, on suspicion that it contained undeclared horse meat. Workers in the plant spoke anonymously to the *Guardian* journalists, stating that they had knowingly labelled horse as beef and that they had also received pallets of old meat that they would be asked to process after regular hours. 'It smelled so bad that we had to cover our face with a cloth.'[4] UK-based meat trader Norwest Foods International Ltd sourced frozen beef from Willy Selten for the beef processor ABP Food Group, owned by Irish beef baron Larry Goodman. ABP's Silvercrest factory in County Monaghan, Ireland made frozen burgers with the beef and sold them on to Tesco, Burger King, Co op and Aldi in the UK. Norwest Foods and Silvercrest claim that they didn't know the beef had been adulterated with horse and pork.

The supply chain took Irish horses to an English abattoir. The horse meat then moved to the Netherlands where it was mixed with beef at Willy Selten. It then travelled back to Ireland through a UK trader where it was processed into burgers and then shipped to the UK for sale. This actually seems pretty straightforward now that the journalists have laid it all out for us. Except that the Silvercrest factory that was making the burgers was using 40 different suppliers and the mixture of meat going into the hopper changes every half hour. Where do you even begin?

While most establishments along the meat supply chain are highly controlled, others are less so. Traders and brokers, for example, are a less regulated but essential step in the food supply chain; they ensure a constant supply of meat gets to the manufacturers and processors for the best price. They help things to run smoothly between manufacturers, processors and retailers and while they may not always be a physical link in the supply chain, they add to the layer of communication and paperwork when trying to trace things back.

Horsegate and dodgy chicken, like all food fraud, requires a certain amount of cooperation among the different points of the supply chain. Whether it's slaughterhouses agreeing to take payment for waste that they previously paid to have removed or workers trimming putrid meat, they are all complicit in the crime. They are not asking questions because they don't really want to know the answers. Professor Elliott's report following Horsegate recognised the critical role industry staff play in identifying crime early on. The report recommends support for further development of whistleblowing mechanisms within industry culture so there is safety in asking questions and reporting unscrupulous activity. It may be reporting boxes of horse trimmings leaving the plant as minced beef or it may simply be questioning prices that are too good to be true. Whether they made a conscious decision or not, the players in these frauds have all helped an illegitimate product to move through a perfectly legal framework.

Though Horsegate and Maggot Pete made a lot of headlines, it's necessary to put meat fraud into some perspective. The USP's Food Fraud Database, which they have kindly made available to the public, includes 25 reports of meat fraud occurring between 1997 and 2012. Compare this with 440 reports of fraud for oil and 320 for milk. Meat is by no means the most commonly tinkered-with food, yet for some reason it gets a lot of attention by the media and makes the public raise their steak knives in anger. It's a sort of 'You can fiddle with my fruit and mess with my milk, but don't cheat my meat' kind of attitude. Perhaps it's because we feel we pay a lot for meat or because it is the centrepiece for many people's meals? Whatever the reason, it's fair to say that a little horse meat in some burgers and lasagne can certainly stir things up. However, it must be said that an unexpected side effect of the horse meat scandal was that the British public gained a new appetite for horse meat. One year after Horsegate broke, sales of horse meat had boomed because out of the media frenzy that ensued, one of the messages consumers took home was that horse meat is lean and healthy. Undoubtedly, too, a number of people were just curious as to whether they could taste the difference.

It's all minced up
We saw in the previous chapter that species substitution is rampant in the seafood industry, but how often do our furry and feathered protein sources get swapped? Was Horsegate the exception or the rule?

In 1995, researchers from the Florida Department of Agriculture and Consumer Services analysed over 900 meat samples collected from Florida retail markets; 806 were raw samples of meat and 96 were cooked. For their analysis they turned to proteomics (the large-scale study of proteins). One of the methods they used was ELISA (Enzyme-Linked ImmunoSorbent Assay), which is based on antibodies recognising and binding to specific animal proteins. For this technique, antibodies are developed for specific proteins, such as a heat-tolerant protein found in the muscle of pigs. The sample of 'all beef' sausage is mixed with the antibody and if the pig protein is present, it will bind to the antibody. The sample is then washed to get rid of any unbound sample. A second antibody can then be added, which is linked to an enzyme. This antibody binds to any bound pig protein. Finally, a labelling reagent is added. This reagent contains a substance that will be converted into a coloured product by the enzyme. This colour change can then be used to simply indicate presence of the protein of interest or it can be measured against a set of standards to determine the concentration of protein present.

Of the 900 meat samples, the researchers found that 149 (16.6 per cent) contained more than 1 per cent of an undeclared meat.[5] The substitution rate was higher among the cooked meats (22.9 per cent) than the raw meats (15.9 per cent). The undeclared species found in minced beef and veal products were sheep, pork and poultry. However, it must be stated that immunoassays will only recognise the species that they have developed and added antibodies for. In other words, unless they added antibodies for rat and dog, they wouldn't have found them.

In 2006, a group of Turkish researchers used immunoassays to test processed meat products such as fermented sausages, salami, frankfurters, pastrami, bacon and canned goods.[6] They found 22 per cent were adulterated; 11 of the 28 sausages that were labelled as beef contained only chicken.

China has been riddled with meat substitution scandals. There have been reports of rat, mink and fox meat being transformed into mutton slices. Twenty thousand tonnes of meat were seized and more than 900 people were detained in association with the scandal. In early 2014, Walmart's operations in China were recalling donkey meat because it had been adulterated with fox meat. Donkey is a very expensive meat and highly sought after for its tenderness and sweetness; fox, not so much.

Pork is swapped for beef, beef is swapped for buffalo, fat trimmings and offal (internal organs) are added to minced beef, chicken is sold as lamb, pork is sold as chicken, and beef and pork gristle and bones are injected into chicken. The list is long, and this is just substitutions between animal species. We haven't yet mentioned the undeclared ingredients – such as added water, chickpea flour, rice flour and soy – that are added to meat to bulk it up.

Of course, there's always the possibility that some of these undeclared species are the result of accidental cross-contamination. When a processor takes a carcass from a slaughterhouse and debones it and takes it down to smaller cuts, there are a number of leftover bits that aren't particularly useful as a cut of meat, and these are called the trimmings. For beef, about 15 to 20 per cent of the carcass will end up as trimmings, so this is a significant amount of meat that it would be shameful to waste. The trimmings are shipped to a processor that then mixes the extra fatty trimmings with the extra lean trimmings to get the desired fat-to-lean content for their customers. It can then be packaged up as mince and sold on to other manufacturers or retailers, or processed further into things like burgers. Processors work with several different types of meat and so there is a possibility that some minced pork remnants will be left in the pipeline and get pushed through when the beef goes through the machine. As a result, there is some forgiveness in levels of contamination. The European Food Safety Authority (EFSA) uses a 1 per cent threshold – anything above this level of contamination is considered to be intentional adulteration.

As with any food, the more processed the meat, the more difficult it is to tell by visual inspection alone whether it's been tampered with. By definition, mince is a mixture of meat that's ground beyond any hints as to its animal origins. The only distinguishing feature the meat sitting in a plastic tray, bound by a thin layer of protective plastic wrapping, has is its colour. We can judge its animal origins based on its shade of red – ranging from pale poultry pink to vibrant venetian red venison. The fat content can be estimated based on the relative proportion of red bits and white bits. Freshness is assessed by the saturation of the colour – is it a dull grey colour or bright red? It's not a lot to go on and even these attributes can be manipulated. While it's easy to distinguish some turkey mince from beef, things can get more difficult between the red meats such as horse and beef.

Products such as sausages are among the most prone to adulteration. While there is the possibility of cross-contamination as we just mentioned, the more cynical (and one could argue realistic) viewpoint is that cheaper substitutions are easier to hide in a processed product. In 1991, researchers from the University of New South Wales, Australia went out and bought samples from butchers and supermarkets of the most commonly consumed sausages – thick beef, thin beef and thick pork.[7] The researchers were interested in nutritional quality, but as there had been an article in the media about adulteration, they decided to test for other species in the samples using the ELISA method as well. Cow, sheep and pig meat were detected in all of the 'all beef' sausages, thick and thin. Of the 10 pork sausages tested, three contained only pork as labelled, three contained undeclared cow meat and the remaining four samples contained cow and sheep in the 'pork' sausage. Of the 30 sausages tested, only the three pork sausages were labelled correctly.

In 2012, researchers in South Africa examined a total of 139 processed meat products – from minced meat to deli meat – to look at what ingredients were not being declared on the label.[8] They used the ELISA method to detect undeclared plant proteins, but also used DNA-based methods to look for a total of 14 animal species. They found undeclared plant and/or animal species in 95 (68 per cent) of the samples. The highest rates of adulteration were in sausages; nearly half of the sausages contained undeclared pork. All together, the meat products tested contained undeclared soy, gluten, beef, water buffalo, sheep, goat, donkey and chicken. The majority of the products were not complying with labelling laws.

Adding or substituting meats and using vegetable fillers to bulk up the end product in sausages and other highly processed meats isn't difficult. It's a matter of adding another ingredient into the giant mixer as it blends together the meats and spices. Doner meat is similar – it's mince and spices mixed together – which is probably why 70 per cent of lamb kebabs from British takeaways tested in 2013 contained cheaper, undeclared meats.

Substitutions aren't limited to highly processed meat

The techniques of the fraudsters are now sophisticated enough that substitutions can happen beyond the minced and processed meats. Let's return to the example from China of fake mutton. Thinly shaved mutton slices are a popular hotpot ingredient. There have now been several scams unveiled in China that have involved the sale of fake mutton; one operation, raided in January 2013, had 40 tonnes of fake mutton and another 540 tonnes of materials to make more. Allegedly, rat, mink, fox and duck meat have all been used as the base for this fake product. These meats are apparently soaked in a cellulose gum (sodium carboxymethyl cellulose), which is commonly used in food manufacturing to extend shelf life, improve freeze/thaw stability and help bind water. In the making of fake mutton, this process allows the meat to take on more water and therefore increase its apparent weight. Food colouring is used to provide the ideal shade of mutton, and food adhesives (more on meat glue later) are used to bind the fake meat with real mutton fat. The end product is a passable, but not indistinguishable, version of mutton. What sets the fake apart is that the fat is not marbled throughout the meat as would be the case naturally. The fat and meat are quite separate and when it is thawed or cooked, the adhesive fails and one is left with fragments of fat and fragments of meat.

Despite what seems like an arduous process, making the fake mutton is worth the effort. The fraudsters can sell it wholesale to restaurants for about £2.12/kg (US$1.45/lb) less than the real thing, allowing them to undercut competitors selling real mutton. Forty tonnes of fake mutton would turn a profit of about £128,000 (US$192,000) – nearly 23 times the average annual salary in China for 2014.

In September 2013, police confiscated 20,000kg (44,000lb) of pork masquerading as beef from a factory in north-west China. Not mince, not sausages, not even thinly sliced 'mutton', but whole cuts of pork that had been made to look like beef. One wouldn't think it was possible. The pork is mixed with beef extract and a glazing agent and left to sit for ninety minutes. When cooked, the meat takes on a dark beef-like appearance rather than the characteristic white pork colour. The beef extracts even give it the beefy aroma one would expect. Though it may be more difficult to swindle people over whole cuts of meat, this shows it's not impossible.

If one end of the spectrum is to transform a pork chop into a steak, the other end is to change things at a microscopic scale. In 2001, the UK FSA released results of an investigation that was carried out jointly with 22 local authorities. They tested 68 samples of chicken breasts that were being sold to the catering trade and found that more than half of them were mislabelled, including some that contained undeclared hydrolysed protein. Hydrolysed protein is protein that's been broken down into smaller segments known as peptides, usually using an enzyme.

This can be a very useful process as it can remove the allergenic properties of proteins and make them more easily digestible. Baby formula, for example, contains hydrolysed milk proteins (casein or whey). Collagen, which is the main structural protein derived from bone, connective tissue, skin and hide, forms the ideal water-retaining agent when it is hydrolysed – gelatin – and it was this that was being added to the chicken breasts.

The protein powder is purchased by processors and made up into a brine solution. This solution is then directly injected into the breasts using needles, or the chicken breasts are tumbled with the solution in a machine like a cement mixer. Either way, the breast meat takes up this solution and the hydrolysed protein helps retain water, even while cooking. The result can be a product that actually contains as little as 55 per cent chicken; the rest is additives, including water. This is a perfectly legal process, but it must be labelled correctly as 'chicken breast fillets with added hydrolysed chicken protein'.

The technique was developed by Dutch processors to introduce protein and water into salted chicken that they were importing from Brazil and Thailand. The processors were taking advantage of an EU tax loophole, as salted meat is subject to much lower import tariffs. By adding water to the chicken, they were making it more palatable but also effectively selling water for the price of chicken. Of the 68 samples taken by the FSA in 2001, 20 per cent contained undeclared hydrolysed protein.

Shortly thereafter, it was revealed that some Dutch manufacturers were not only adding undeclared hydrolysed protein, but also the protein was being extracted from other animals. The FSA conducted DNA testing on 25 samples and found that almost half of them contained traces of DNA from pigs, though all but one of those samples were labelled as halal (meat that adheres to Islamic law and certainly would not include pork). They suspected that beef protein was also being used, but their DNA-based methods weren't picking up any beef DNA. The hydrolysed protein powders are extremely processed, making any DNA, if present at all, very difficult to detect – particularly when looking for a very small amount of beef or pork DNA in a lot of chicken. The proteins are also fragmented through processing, which eliminated the use of immunoassays, such as ELISA. The FSA needed a new test.

The FSA collaborated with researchers from the University of York who had developed new procedures to identify species of ancient bone fragments dug up in archaeological sites. Archaeological work suffers the same challenges faced by food forensics in that the proteins have decayed – though in the case of archaeology it is through time rather than processing. The researchers had discovered that the collagen protein found in bone has enough variation between species to be useful in fingerprinting collagen-based tissues (for example, bone, cartilage, skin, tendon, blood vessels). Luckily, the hydrolysed protein in the chicken breasts had been extracted from these types of tissues.

The technique the York researchers have developed is called ZooMS, short for ZooArchaeology by Mass Spectrometry. For the analysis, proteins in the tissue sample are cut up into peptide fragments using the enzyme trypsin. The mass of each peptide is then determined using time-of-flight mass spectrometry, which essentially shoots the peptides out using an electric field and uses a detector to see how fast they fly a particular distance. This provides a unique mass-to-charge ratio for the peptides. Certain peptides (fragments of the collagen protein) are species-specific and can be identified by comparing them with a library of collagen proteins developed for different species.

As the chicken processors don't make the hydrolysed protein powders themselves, the FSA conducted an investigation of the powders directly.[9] They obtained five sample powders, four of which were made in the UK and all of which were labelled as containing only poultry-derived protein. They ran them through numerous tests, including the methods developed at York. They used real-time PCR to look for chicken DNA in three of the powders and found that two powders tested positive for chicken DNA only and one tested positive for chicken and pork DNA. Had the tests ended there, one might have suspected that only one of the powders had used species other than chicken. Luckily they didn't stop. Analysis of the collagen protein showed that *none* of the protein in all five powders had been derived from chicken. All of them contained bovine collagen-specific peptides and two of them contained bovine and porcine-specific peptides. Interestingly, two of the powders also contained unidentified non-food animal peptides. The powders had tested positive for chicken DNA probably because a small amount of chicken blood had been added, which would mask any pork or beef DNA that was likely to be highly degraded. It was the analysis of the collagen proteins themselves that revealed the true sources of the hydrolysed protein in the powders. Mislabelled powder means that some chicken processors may be unaware that they're injecting protein from other species. Some processors have shifted to the use of plant-based protein powders as a result. Yet it does not eliminate the fact that the process is introducing a lot of water that consumers are paying for, which is fine as long as it's labelled as an added ingredient. Consumers can then make their own decisions about whether they want to pay for water. But when it's undeclared ... that's fraud.

Shaped shanks

Just as we consumers want the ease of a deboned, skinned fillet of fish, we are also on the lookout for no-mess meats. This, along with the industry's genuine desire to reduce waste, has led to the evolution of formed meats. This perfectly legitimate process has found itself on the shady side of food fraud more than once. Ham is a perfectly good example of a popular formed meat. Traditionally, a ham referred to the rear leg of a pig, soaked in brine, smoked or dry-cured, then hung for weeks under the protection of a layer of fat and skin. These days, we can't afford the time to hang our hocks and who wants to be slicing through a leg in the morning to make school lunches? We want our ham to be lean, pristine and perfectly shaped to fit within the confines of two slices of bread. Therefore, the majority of ham we find in supermarkets is formed from a number of different muscle cuts. The cuts of meat are sorted into batches of similar colours so that the end product looks like a product from a single muscle. The cuts are mixed with phosphates, salt, nitrite, colour enhancers and water. These additives give the end product the right colour and flavour and also help release the protein myosin from the meat, which acts as a natural gelling agent to help fuse the pieces together. The meat cuts are placed into a cooking bag and then into a meat press and cooked. The result is a nicely formed ham that can be sliced up and packaged ready for sandwich making.

Where the process starts to enter shady territory is when added ingredients, such as water, are not declared on the label. The process of forming and curing ham requires the introduction of water. But a report in 2005 by consumer watchdog Which? showed that wafer-thin hams from the major supermarkets contained up to 25 per cent water. Samples of canned ham were even higher. In response to the report, some manufacturers stated that they were offering products within a range of budgets and in order to produce cheaper products, more water is added. This is perfectly legal. In the UK, ham is understood by consumers to mean a formed product that contains water, but if more than 5 per cent of the product weight is added water, it needs to be declared on the label. Yet, there isn't any requirement to say how much water is in the product. It's easy to see the path to the dark side.

While ham is clearly something consumers recognise as a formed product, food technologies have become so advanced that some formed and reformed meat can be far less recognisable. Binding together several cuts of meat to form a single piece that, for most people, is indistinguishable from a cut that originates from a single muscle, such as filet mignon, has become a common culinary technique. The cuts used to form the meat aren't necessarily of lower quality, but are perhaps just too small to form a single portion on their own – so it shouldn't be assumed that the formed meat is of poor quality. However, once again, it needs to be declared as formed or reformed meat so as not to deceive customers. Labelling laws in most countries require formed or reformed meat to be labelled as such, but restaurants and the catering industry aren't required to do so.

The meat cuts are bound together by transglutaminase. This is an enzyme that's essential to life in that it catalyses the bonding between proteins. In a healthy organism, transglutaminase helps form the biological polymers important in blood clots as well as the growth of hair and skin. Elevated levels of the enzyme, however, have been associated with neurological diseases such as Huntington's and Parkinson's and it may be responsible for helping to form the protein aggregations associated with these diseases. Transglutaminase is most commonly produced in commercial quantities through natural fermentation of the bacterial species *Streptoverticillium mobaraense* or through extraction from animal blood. It has been dubbed 'meat glue' in some circles and has received considerable attention from the media in North America. However, before we discuss the ethics of meat glue, let's first take a moment to discuss blood-derived products in general as it may help inform the discussion.

Many millions of tonnes of blood are released (for lack of a better word) each year through the slaughter of animals for food. Blood is very high in protein (about 18 per cent). In fact by weight, blood is a better protein source than eggs and most meat. Many cultures have traditionally turned blood into culinary products, such as black pudding or sausage (eaten in a number of European countries, including Britain), blood pancakes (many Scandinavian and Baltic countries) and even blood tofu (China) to take advantage of this protein-rich resource. Yet many countries, including the US and Canada, are generally squeamish about blood-based food and so much of it is either literally poured down a drain or disposed of in a more responsible manner. This is incredibly wasteful and if a life must be taken for food it seems only respectful to use as much of it as possible. Industry profits are also greatly improved by reducing waste, which is why there was an outpouring of research into the potential uses of blood in the 1990s. By 2001, the industry was using about 30 per cent of blood released in slaughterhouses, and by now it is likely to be much more. Blood proteins are used by the food industry as binders, natural colour enhancers, emulsifiers, fat replacers and meat curing agents. For example, the red blood pigments are isolated and used as a natural colourant to enhance and homogenise the colour in that formed ham we referred to previously or, more dishonestly, to colour pork with a beef-like hue. These natural colourants are also used to increase the contrast between fat and meat so that salami isn't simply different shades of grey.

Other blood products take advantage of the gelling properties of plasma. Plasma is tasteless and colourless and is a better binding agent than egg white – not to mention being less expensive. Food additives derived from blood plasma and isolated blood proteins, such as transglutaminase, duplicate the natural bonding process of muscle tissue and therefore make fabulous meat glue.

The point is, things aren't always black pudding or white. Products such as meat glue are innovative responses to the waste associated with the food industry. Meat glue is a natural product that makes use of an otherwise wasted product (blood), and it's used to transform a less valuable product (small cuts of meat) into a better quality product (a perfectly portioned piece of meat). In the right hands, it is a tool that can be used to create a culinary delight. In the wrong hands, it is an agent of deceit.

In 2010, the European Parliament decided that the risk of misleading consumers was too great and turned down the use of any additives that can be used as meat glue. They felt the benefits of such a product were outweighed by the risk of deception. EFSA had provided a positive opinion on meat glue in 2005 in terms of its safety. But the European Parliament's 2010 decision took into consideration concerns that fusing pieces of meat together placed the outside surface areas of the meat, which are most prone to bacterial contamination, on the inside of the formed meat. As we all know, the inner portion of meat isn't exposed to the same high temperatures as the surface, which may present a risk in terms of food safety. Alas, no meat glue for Europe.

Other mischievous meat manipulations

There are other forms of meat fraud in addition to disguising rotten meat, substituting species, adding undeclared water and mislabelling formed meat. Cheaper cuts of meat can be labelled as more expensive cuts – a rump steak masquerading as a sirloin, for example. Detecting this type of fraud comes down to superior product knowledge.

False claims can be made about the animal's diet (grain-fed versus grass-fed) that can help fetch a higher price for the product. Meat animals can be reared to a slaughter weight much faster on a grain diet, containing corn and soy, than on a grass-only diet. However, grass-fed animals are sold for a premium because of some publicised nutritional benefits over those fed on grain. Meat from grass-fed animals is reported to be much lower in calories (though some argue that this diminishes the taste), to contain more omega-3 fatty acids and to have up to seven times more beta-carotene than meat from grain-fed animals. Once again, there is a price differential between two products that are indistinguishable to the consumer, presenting an opportunity for fraud. Grass-fed versus grain-fed can be distinguished using the same principles used to reveal corn syrup in honey and adulteration of maize oil. Corn, you may recall, is a C_4 plant and a main constituent in grain diets. Therefore, the carbon isotope ratio of the meat can be used to distinguish whether the animal was eating mostly corn or mostly C_3 grasses.

Chickens can be dishonestly labelled as free-range in order to earn a premium. Free-range chickens are more expensive than barn-reared, in part because they are raised in lower densities and slaughtered older than barn-reared. The premium for free-range chickens helps offset the fact that farmers can move fewer birds through their premises annually. And conscientious consumers don't mind paying a little more for these birds as they believe that they have had a marginally better life. Once again, stable isotope analyses can be used to authenticate such claims. However, it is the nitrogen isotopes that are key here. Chickens that have had access to the outdoors naturally get more meat in their diet as they consume insects and other invertebrates outside. Because the heavier ^{15}N isotope has a higher retention level in tissues, this is passed along the chain as animals eat each other. Animals higher in the food chain will accumulate more ^{15}N relative to ^{14}N in their tissues, which gives free-range chickens a different nitrogen isotope ratio from those raised on grain indoors.

There can also be deception around the geographical origins of the animal. Such dishonest labelling may be to avoid tariffs or contravene import bans. In some cases, it may be to appeal to consumers who are trying to buy products produced in their own country – customers trying to 'Buy British' for example. Country-of-origin testing is perhaps one of the most challenging areas of authenticity testing. The isoscapes described in Chapter 2 can help to reveal the origins of meat, because luckily we are what we eat as well as where we eat. The isotope ratios of oxygen, hydrogen, carbon, nitrogen, sulphur and strontium can be used together, along with measurements of trace elements, to narrow down where in the world an animal was reared. Isoscapes are useful tests to confirm that a lamb roast, for example, is consistent with an isotopic signature from the British Isles. They are less useful in taking a sample of meat and asking where in the world it is from, but this will change as more baseline data are gathered and collated from around the world.

Labelling laws are changing to try and improve traceability of meat products. Country-of-origin labelling has existed for beef in Europe since 2002, but as of April 2015 all fresh, chilled and frozen pork, sheep, goat and poultry also has to be labelled with country of origin and country of slaughter. And in February 2015, the European Parliament backed a resolution to introduce mandatory country-of-origin labelling for any meat used in processed food.

Motivations and implications

Though one might assume that the tight regulations around meat would discourage the fraudster, we have seen that this is clearly not the case. And, at the risk of being repetitive, the motives behind these meat shams are consistent with other types of food fraud: it's about money. Again, the mindset of the criminals involved may fall anywhere along a spectrum from blind greed to innovative problem-solving. The swindlers may be opportunistic or systematic in their approach. But in the end, someone stands to gain from the action.

The implications of meat fraud are also similar to those we see in other commodities. There are obvious economic costs – to governments, to the industry and to consumers. Just as with seafood substitutions, we see mislabelling of species in order to avoid costly tariffs put in place by governments. The industry faces product recall costs, potential legal action and a hit to their bottom line as consumer trust is shattered. Consumers are out of pocket when they pay for one thing and get a lesser product ... even, in some cases, just water.

There are also severe health implications. It is only a matter of time before serious illnesses are caused by eating condemned meat that has been recycled back into the human food chain. With Horsegate, health concerns were raised that the animals were never intended for human consumption and therefore may have contained residues of drugs (steroids and antibiotics) beyond any legal limits for food. Substitutions may also mean that there are undeclared allergens that present a health risk to consumers.

Deception around the geographical origin of meat can have health risks for humans and other animals as well as other consequences. Concerns over BSE and vCJD prompted the EU to impose compulsory beef labelling rules in 2000 that require the geographical origin of the beef to be clearly labelled. Yet the beef protein injected into chicken breasts was of unknown provenance and therefore not without risk, as the prions responsible for these diseases are incredibly resistant to heat and processing. Furthermore, meat coming in from regions with a history of food-borne pathogens that is labelled as coming from a different origin may not be subjected to the same testing.

As with seafood substitution, there are environmental costs associated with meat substitution. An authenticity study carried out on South African wild meat products in 2013 found mountain zebra (*Equus zebra*), which is on the IUCN Red List, being sold as kudu (antelope).[10]

Perhaps relatively unique to the meat industry, however, is that there are numerous groups whose values and beliefs dictate what meat, if any, they eat and how it should be killed. Islamic and Jewish dietary laws prevent the consumption of pork. Therefore, the addition of pork in a lamb curry or a sausage labelled as all beef has serious religious implications. Hindus don't eat beef, and so the knowledge that hydrolysed beef protein may be found in chicken breasts (labelled or not) is no doubt very distressing.

Vegetarians and vegans are, of course, not likely to be victims of meat fraud, and yet they aren't entirely immune to animal-based swindles either. You will read in Chapter 7 that saffron has been adulterated with meat fibres. In 2014, routine government testing in Malaysia found pork DNA in two chocolate bars produced by Cadbury's. This is concerning not only to sweet-toothed vegetarians, but also to the Muslims who make up more than 60 per cent of Malaysia's population. Cheap blood protein is being explored for its emulsifying and coagulating properties outside of the meat industry. Blood proteins have been used to replace the very costly eggs used in many baked goods. Regulations state that this would need to be declared on the label, but catering, restaurants and bakery items sold loose don't carry labels and one might not think to ask whether a cake contains beef protein. Though it seems somewhat counter-intuitive, vegan products can be more expensive as plant-based products are used instead of dairy foods; the premium price creates the economic incentive to substitute some cheaper ingredients.

Thank goodness for a little horse

As well as generating a market for horse meat in the UK, the Horsegate scandal managed to enrage consumers around the world who wanted greater transparency and more assurance of authenticity in their meat products. It has been a catalyst for potentially long-term and meaningful change.

The UK government responded with the Elliott review, which took an in-depth look at the integrity of Britain's food supply. The findings were alarming and the recommendations substantial. The UK government accepted it all and started to implement recommendations outlined in the report, including the formation of a dedicated Food Crime Unit.

Businesses are now conducting fraud risk assessments, mapping their supply chains beyond the links connected directly to them (one step forward, one step back). They are taking a hard look at where their vulnerabilities lie: these may be on the floor of the processing plant, in their supply chains and/or in their offices. As we write this chapter, the trials linked to Horsegate continue to unfold and it is evident that food fraud doesn't happen as an isolated offence. Books have to be cooked and paperwork needs to be forged. The Dutch meat wholesaler involved in Horsegate, Willy Selten, was found guilty of forging invoices, labels and decorations, and using forged documents to sell meat. He was sentenced to two and a half years in jail – a lenient sentence as the judge felt that Selten, now bankrupt and facing damage claims of €11 million (almost £8 million, over US$12 million), had already been punished. It's the paperwork that gets them in the end.

Consumers in the UK changed their buying habits, claiming to buy less processed meat, fewer ready-meals and avoiding brands involved in the scandal. Immediately following the horse meat scandal 19 per cent of consumers surveyed said they were avoiding the products involved. Yet, one year later, this dropped to 9 per cent. We're a forgetful and forgiving bunch.

What was ignored in the horse meat scandal was our relationship with the animals we eat. The fact that there was also pork in the minced beef didn't make it into most of the media, but because we have a different relationship with horses from that with pigs, the horse meat made headlines. This touches on perhaps the most widespread fraud in our society, and we use the word fraud in this sense not in terms of a criminal act but in its reference to deceit. It is this. When we put down a beloved cat, dog or horse, we do it with the greatest of care and thought for the well-being of the animal. We are there with the veterinarian as an injection is administered to relieve our loved one of its pain and we pat a nose, hold a paw or stroke an ear as they pass. Unless we are raised on farms, or have seen the inside of a slaughterhouse processing up to 2,000 birds per hour, this is our understanding of animal death. And it is with this misconception of death that we so freely consume animals.

We won't repeat what we've said elsewhere about reducing vulnerability to fraud – it's the same with all commodities. When it comes to meat, however, there are a couple of additional things to keep in mind. Look at labels for information about added water, hydrolysed protein and formed or reformed meat to help you with your purchasing decision. If you choose to buy meat, buy less but of higher quality – there will be some nutritional and environmental benefits associated with this as well.

As for the future of meat fraud, we are looking at a growing population of meat lovers and it is evident that something will have to give. Thomas Robert Malthus, a great scholar and contemporary of Frederick Accum, said 'Population, when unchecked, increases in a geometrical ratio. Subsistence increases only in an arithmetical ratio. A slight acquaintance with numbers will show the immensity of the first power in comparison with the second.'[11] There is less land available to rear more animals that require more food and the energetics of it simply don't make sense. Of all the grain used by developed countries, 70 per cent is fed to animals, which then use 90 per cent of the energy from that grain to keep themselves warm. With every step of the food chain there is considerable energy loss and so it is more efficient to eat low on the food chain. Add to this the energy inputs that modern-day farming requires – from irrigation and fertilisers to the energy required to convert ecosystems into pasture land – and the numbers behind a largely animal-based diet are not favourable. Remarkably, we have looked towards engineered muscles as a potential solution – meat grown from stem cells and exercised into roast-sized splendour independent of any living creatures. If this is our future, consider the potential for fraudsters. With some engineered muscle, horse trimmings, a little meat glue and some hydrolysed protein, the possibilities for deception are endless.

CHAPTER SIX

Milking It

In the mid-1850s New York City found itself at the centre of one of the most abominable food fraud and food safety scandals ever perpetrated: the 'swill milk' scandal. It all revolved around swill (or slop) milk, which came from cows that were housed adjacent to the city's many distilleries and breweries and were fed the leftover mash, called swill, from the brewing process. A *New York Times* report in 1853 and an exposé in May 1858 by the news magazine *Frank Leslie's Illustrated Newspaper* drew attention to the awful conditions of the so-called distillery-dairies of Manhattan and Brooklyn. The cows were tied up in dark narrow stalls in their hundreds, sometimes thousands. Inevitably, they would end up standing in their own excrement, covered in flies and suffering from a range of diseases. The swill milk being produced in these distillery-dairies was described in a *New York Times* editorial on 13 May 1858 as a 'bluish, white compound of true milk, pus and dirty water, which, on standing, deposits a yellowish, brown sediment that is manufactured in the stables attached to large distilleries by running the refuse distillery slops through the udders of dying cows and over the unwashed hands of milkers ...'. The swill milk was then reportedly mixed with water, eggs, flour, burnt sugar and other ingredients to increase the volume and mask the adulteration. Swill milk was then falsely marketed as 'pure country milk' or 'Orange County Milk'.

The *New York Times* editorial, entitled 'How We Poison Our Children', made some bold statements. It reported that Dr A. K. Gardner of the New York Academy of Medicine had provided the results of analyses that suggested a connection between swill milk and high levels of infant

Figure 6.1. 'Swill milk' tethered in its stall in one of New York's distillery dairies in the mid-1800s.

mortality in the city. They claimed that 8,000 children had died in 1857 owing to swill milk. Although these accusations were dismissed by the distillery owners, the *New York Times* journalist, John Mullaly, had also uncovered that 120,000 quarts (136,383 litres) of milk labelled as 'country milk' was delivered to New York residents each day, yet only 90,000 quarts (102,287 litres) was entering the city from the neighbouring countryside. Public opinion ultimately prevailed. The swill milk scandal led to an appraisal of both New York's milk industry and a wide range of other food quality and food-related public hygiene issues.

Our special relationship with milk

As mammals, we have a very particular relationship with milk. It is a critical component of our diet as we develop from vulnerable infants and this is perhaps what makes it so shocking when there is malpractice or adulteration surrounding this fundamental food. Milk is a unique substance nutritionally: the perfect food produced by mammals to sustain their newborn young. Yet somewhere in our history we broke with mammalian tradition and began incorporating milk into our diet well beyond weaning age. And more oddly still, we began drinking the milk of other species! We can only guess at what motivated our ancestors to take this step. Perhaps it was desperation. Perhaps it was the realisation that nutrition can be gained from the animal's flesh or its milk and one is clearly more finite than the other. Or perhaps it was the more contemplated conclusion that if babies grow so fast drinking milk, this must surely be an advantage in adults as well. Regardless of how or why, our ancestors eventually recognised the nutritional advantages of consuming dairy products from their recently tamed sheep, goats and cows. They began developing complex farming practices that eventually led us to rely on milk as part of our everyday diet.

Exactly when humans began to consume the milk of other animals is unknown. It appears from the archaeological evidence, including food residues in ancient cooking pots and animal bones of dairy herds,[1,2] that the practice probably dates back to the very origins of agriculture, more than 10,000 years ago. The journey from early prehistoric farmer to a twenty-first-century milk-consuming society has involved a complex sequence of events and some fundamental changes in both the animals that produce it and the humans that consume it.

While we can all consume milk in early life, the ability of adult humans to digest milk is a relatively recent evolutionary development. The normal state for all mammals, including humans, is to lose the capacity to digest milk after infancy. Our ancestral farmers would not initially have been adapted to drinking milk as adults;

however, as milk drinking became widespread, an evolutionary change occurred that allowed many people to drink milk throughout their lives. The enzyme lactase is needed to break down the milk sugar lactose. The gene that codes for this enzyme is switched on in infants and then is turned off after weaning. A mutation occurred, mostly in European populations, that kept this gene switched on into adulthood. People with this mutation were at an advantage, with researchers estimating that they may have produced up to 19 per cent more fertile offspring than their peers. It became possible for humans to incorporate milk into their everyday diet.[3] Today, about 35 per cent of the human population can digest lactose into adulthood, and most of these have European ancestry, though there are other lactase hotspots in West Africa, the Middle East and South Asia.

Without the enzyme lactase, people are said to be lactose intolerant or lactase non-persistent. Lactose intolerance is an inability to digest the milk sugar lactose, which can cause abdominal bloating and cramps, flatulence, diarrhoea, nausea, rumbling stomach and even vomiting. Those early lactose-intolerant farmers would potentially have been quite ill if they had consumed large amounts of milk from their animals. However, they began to get around this by adopting processing techniques that remove or reduce the lactose concentration to non-toxic levels.

Reducing the lactose content can be done in two main ways. The lactose can be physically separated, such as in butter or cheese making, where the lactose is left in the liquid whey, or it can be reduced through fermentation, such as in yogurt. It seems as though our ancestors had hints of this, as evidence of milk processing and cheese making dates back to the earliest stages of milk consumption more than 7,000 years ago.[4]

There is a huge cultural influence on the types of milk products made and consumed, creating enormous diversity globally. There are 1,000 different types of cheese produced in the world, with France alone accounting for 400 of these. As we learned with olive oil, increasing the different types or classes of a food can increase the potential for fraud, particularly when there are large price differentials, as there can be among the mind-boggling diversity of milk products. Within one UK supermarket the price for cheese ranges from £2.45/kg (US$8.36/lb) for a soft (cream) cheese to £21/kg (US$71.68/lb) for a Parmesan. Add to this the fact that while the vast majority of milk is produced from cows, milk can also be produced commercially from other animals, including buffalo, goat, sheep, camel, donkey, horse, reindeer, yak and even moose. Combined, these account for only around 15 per cent of the global production of milk. Though not necessarily sold in quantity, they may carry a premium; cheese made from donkey milk in Serbia, known as pule, is one of the most expensive dairy products in the world, sometimes selling for as much as £176/kg (US$600/lb). Clearly, including the milk of different animals expands the range of speciality or gourmet milk products still further.

Billions of people consume milk or milk products in some form on a regular basis. As well as pouring milk over our cereal and frothing it into our lattes, we use milk and milk products in a large amount of our cooking and confectionery. Global annual milk production currently stands at 800 million tonnes coming from around 1.5 billion cows; this is a 60 per cent growth in production since the early 1980s. This growth is largely due to an expansion in production (and consumption) in South Asia and, to a lesser extent, Africa. Dairy products are a vitally important component of agricultural production in the developed and developing world. In the developing world animal products, such as milk and other dairy products, are of great importance in preventing malnutrition by providing a vital source of high-quality protein.[5]

The high production levels and globalisation of the dairy industry has pushed prices down. This inevitably squeezes profit margins, a pattern which can lead to fraudulent activities to protect profits. Next to olive oil, milk is the next biggest target of food fraud, constituting just over 24 per cent of reported food fraud incidents in the last 30 years. Fraud in the dairy industry includes increasing water content, reducing fat content, misrepresenting species origin or country of origin, addition of non-dairy protein, addition of vegetable or animal fats to milk fat and wrongful addition of a wide range of other ingredients – from detergent to formaldehyde – to various dairy products, including infant formula. Other activities include the fraudulent addition of whey to other dairy products or the inappropriate use of certain dairy processing technologies, such as membrane filtration to remove high-value milk protein fractions. The diversity of dairy products requires tight regulations on product labelling. The detailed legislation on dairy foods has required extensive standardisation of testing methods, which have been compiled by the Association of Analytical Chemists, the European Commission and the International Dairy Federation. Yet, despite tight regulations and standardised testing methods, criminals are messing with our milk.[6]

What is milk made of?

Milk has an extremely complex chemical composition, which luckily is of considerable value in detecting fraud. The beautiful opaque white colour of milk comes from the emulsion (or colloid) of milk fat globules in water, together with dissolved carbohydrates, protein aggregates and minerals. Water is the main constituent of milk – about 87 per cent in cow's milk. The milk fat (butter) is a mixture of triacylglycerol, with three fatty acid molecules linked to glycerol as you may recall from the oils in Chapter 3. The number of carbon atoms in the fatty acids in butter ranges from 4 to 18. This differs from the composition of ruminant body fat and vegetable oil, which generally range between 14 and 18 carbon atoms. This gives rise to a much more complex triacylglycerol composition in milk fat; at least 120 different triacylglycerol are present, while in animal body fat and vegetable oil there are only around 40 different triacylglycerol.[7, 8, 9] The triacylglycerol fat globules are surrounded by a thin layer or skin made up of a related group of compounds, known as phospholipids, and proteins. This layer works as an emulsifying agent, keeping the individual globules separate and suspended in the water. The fat globules also contain low concentrations of diacylglycerols and monoacylglycerols, free cholesterol and cholesteryl esters (a dietary lipid), free fatty acids and fat soluble vitamins A, D, E and K.

Normal bovine milk contains 30–35 grams of protein per litre (1.9–2.2oz/pint), of which around 80 per cent is the casein proteins. The water-soluble whey proteins, mainly lactoglobulin, make up the remaining 20 per cent of protein in milk. Unfortunately, many people have allergies to one or more of the many milk proteins. A milk allergy is different from lactose intolerance. While the effects of lactose intolerance (being unable to digest milk sugars) are brought on by the consumption of larger amounts of milk, an allergic reaction to milk protein is triggered by the consumption of any milk. If someone is diagnosed as having a genuine milk allergy, the recommendation is total avoidance of milk proteins.

The whey (what's left over when you take the fat and caseins out) also contains carbohydrates. The main carbohydrate is lactose (around 5 per cent), which is made up of the two sugars glucose and galactose. The carbohydrates remain in the whey, together with the whey proteins, when the curds are coagulated, as in cheese making. Milk also contains a number of important minerals, including calcium, phosphorus, magnesium, sodium, potassium and chloride. Furthermore, it's rich in vitamins B6, B12, C, thiamine, niacin, biotin, riboflavin, folates and pantothenic acid. Cow's milk also contains white blood cells, mammary gland cells, various bacteria and enzymes. The bottom line is that milk is a highly complex 'soup' of nutrients, so any fraudulent activity is likely to have significant health impacts on individuals who rely on it.

In search of dairy substitutes

So widespread are our society's milk product-based eating habits that we have gone to great lengths to develop a wide range of milk and dairy product substitutes that are capable of filling this niche for those who avoid these animal products – whether through personal choice or adverse health reactions to milk. While these products are wonderful legitimate alternatives to milk, they also illustrate that a white liquid substance can be copied relatively easily.

The most obvious of the milk product substitutes are the plant 'milks', such as soy, almond, rice, coconut and hemp milk. Other alternatives are appearing on the market, including quinoa milk, oat milk, potato milk, 7-grain milk (from oats, rice, wheat, barley, triticale, spelt and millet) and sunflower milk. However, as indicated above, there's much more to milk than its colour, taste and texture, and while some of these milk substitutes might seem perfectly palatable, milk plays such an important role in meeting our nutritional needs that milk substitutes can probably never provide a complete replacement for milk's exceptional nutritional properties. However, they can, if properly matched to the nutritional profile of dairy milk and fortified with essential minerals and vitamins, be a perfectly acceptable part of a diet containing other whole foods.

The origins of genuine dairy substitutions can provide insight and context for dairy fraud. Whether spurred on by dietary requirements or economic hardships, the methods used in the development of some of these products have also been utilised by fraudsters – the difference, of course, being what's declared on the label.

Butter, arguably one of the most wonderful products of milk, is in such demand that butter substitutes are very common. Real butter is made by the simple churning of fresh or fermented cream or milk, to separate the butterfat from the buttermilk. The butter produced is an emulsion of fat, protein and water. We all recognise butter's soft pale yellow colour, which comes from the plant pigments in the animals' diet. However, butter can be pale – almost white – so food colourings are sometimes added during manufacturing to enhance the colour. One of the less desirable properties of butter is that it gets hard in the refrigerator, so being able to spread it on morning toast requires some forward planning. The most familiar butter substitutes are the margarines or related 'spreads', which look like butter, but must be properly labelled as such.

The search for a butter substitute began in the 1860s when Emperor Napoleon III of France offered a prize to anyone who could make a satisfactory butter 'copy', suitable for use by his army and the working classes. Napoleon wanted a cheaper version as edible fats were in short supply in Europe at the time. A French chemist, Hippolyte Mège-Mouriès, came up with the idea of mixing beef fat with skimmed milk to give a cheap butter substitute, which he called 'oleomargarine'. This name later became shortened to the familiar trade name 'margarine'. But this was only the beginning of the margarine story. The aim was still to produce a cheap butter substitute, so reducing the milk content would reduce the cost even further. In 1871, Henry W. Bradley from New York State patented a process for creating what he described as a 'new and improved article of manufacture, a lard, vegetable butter, or shortening', which involved steam-treating a blend of animal fats and vegetable oil.[10]

The next major advance in butter substitutes came with the discovery of vegetable oil 'hydrogenation'.[11] In this process, hydrogen gas is bubbled through the vegetable oil in the presence of a metal catalyst (nickel) at about 60°C (140°F). The result is frighteningly simple and effective: hardened oil. As discussed in Chapter 3, vegetable oils are made up of three fatty acids linked to a glycerol backbone to make triacylglycerol. Although the same basic structure exists for animal fats, including butter, the fatty acids in vegetable oils are more unsaturated (they have high numbers of carbon-carbon double bonds, $C=C$), while in animal fats they are mostly 'saturated' (they have fewer $C=C$ bonds and more single $C-C$ bonds). This is what makes vegetable oils liquid and animal fats solid. The hydrogenation process adds hydrogen to the 'unsaturated' fatty acids, turning them into 'saturated' fatty acids and making them hard at room temperature, just like animal fats. In Figure 6.2, you can count the number of $C=C$ bonds in the triacylglycerol molecule before hydrogenation (seven) and after hydrogenation (two) – the fewer the double bonds, the harder the fat will be.

The introduction of hydrogenated vegetable oils meant that butter substitutes could be produced from cheap vegetable oils without the use of animal products of any sort. This helped get around all sorts of supply problems during times of shortage, such as during World War II rationing. It provided those avoiding animal products with an alternative spread and it was perceived as a healthier choice for those trying to reduce their consumption of saturated animal fats. By the 1950s no margarines containing animal fat of any sort were being produced. However, this did not halt developments aimed at a better margarine or 'spread'. The challenge in the 1960s became one of producing a spreadable butter substitute that actually tasted like butter! The only way of doing this seemed to be by incorporating some milk products back into the vegetable oil spreads that had proliferated in the market.

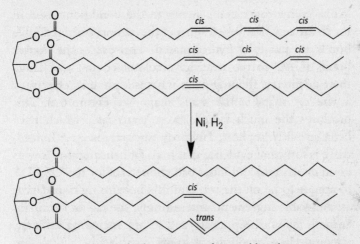

Figure 6.2 Catalytic hydrogenation of vegetable oil in margarine making reduces the number of double bonds which would produce a hard fat. Note also the change to a trans shape for one of the double bonds in the lower triacylglycerol.

Making margarine industrially currently involves taking vegetable oils and fats and chemically modifying their physico-chemical properties using fractionation (separation), interesterification (moving fatty acids between triacylglycerol molecules) and/or hydrogenation. The modified oils and fats are then mixed with skimmed milk and the mixture emulsified, chilled to solidify and whisked or stirred to improve the texture. Alternatively, if no solid fats are added to the vegetable oils, full or partial hydrogenation is necessary to harden the oil. The hardened oil is then mixed with water, citric acid, colouring, vitamins and milk powder, lecithin (emulsifier) to help keep the added water evenly distributed throughout the oil, salt and preservatives.

A factor having serious effects on the margarine market and related parts of the food industry is that the ingenious and seemingly innocuous hydrogenation process appears to produce damaging health effects. When vegetable oils are 'partially hydrogenated', owing to the conditions used in the hydrogenation vessel, some of the carbon-carbon double bonds are not fully hydrogenated. This causes the double bonds (C=C in the triacylglycerol molecules) to change their geometry, flipping from what is known in chemistry as the 'cis' shape to the 'trans' shape (see Figure 6.2). This produces the much talked about 'trans fats', which have been seriously implicated in cardiovascular disease.[12] Indeed, there is sufficient evidence that trans fat consumption lowers good cholesterol and raises bad cholesterol for the US FDA to appear to be on the verge of finally banning trans fats as an allowable ingredient. Interestingly, the fats of ruminant animals (cows, sheep, goats, and so on), including milk fat, contain trace amounts of trans fatty acids, which are produced in their digestive system through a natural bio-hydrogenation process, although the amounts consumed from this source are regarded as minor.

And so was the rise and fall of margarine. It was developed in response to animal fat supply shortages in the nineteenth century and didn't truly become popular until nearly a century later, when people began avoiding animal fats for perceived health reasons. It then fell out of favour 60 years later as federal governments – the very same ones that supported its original development – cracked down on trans fatty acids. It is yet another example of the complicated road we pave when we process foods.

I can't believe it's not butter

Milk fat, hence butter or ghee (a type of clarified butter that's produced by evaporating the water from butter), is the most valued component of milk. This makes it an obvious target for adulteration or substitution. As the market for butter substitutes grew and processing methods such as hydrogenation developed, the temptation just became too much for the fraudsters. Thankfully, the adulteration of butter is relatively easy to detect, largely on account of its distinctive chemistry.

Mixing vegetable oil into butter is the most likely adulteration. However, it's extremely easy to determine this by performing a sterol analysis using GC or HPLC. Sterols are common compounds in plants and animals; they are major components of cell membranes and they are precursors to steroid hormones. Animals only produce the sterol cholesterol, while the plants used for vegetable oils produce the closely related sterol, sitosterol (see Figure 6.3). Therefore, the presence of sitosterol can be used to determine whether animal-based fat (butter) has been adulterated with an undeclared vegetable-based fat (oil). Though it may seem difficult to spot the difference between these two compounds, the branch side-chain on sitosterol means it is heavier than cholesterol and it will appear at a later time in a gas chromatogram, giving a different peak on the chromatogram from cholesterol. Plants produce other sterols that may also be found together with sitosterol, including stigmasterol and campesterol (together known as phytosterols). The presence of any of these compounds in butter is definitive proof of its adulteration and this analysis is a European Commission (EC) approved procedure for butter authentication.

A recent survey of a range of butters from Polish supermarkets illustrated the power of these approaches in detecting butter fraud.[13] Of 16 butters investigated, two showed detectable sitosterol, indicating a vegetable oil component. This was supported by an analysis of the triacylglycerols, which showed an abnormally high concentration of polyunsaturated fatty acids, which was consistent with vegetable oil being added. Sterol analysis cannot, however, be used to detect the addition of animal carcass fats to butter as all animal fats contain cholesterol. As we mentioned earlier in this chapter, there are differences in the distribution of fatty acids in the triacylgycerols in milk fat versus body fat. These differences can therefore be used to detect the addition of other animal fats to butter.

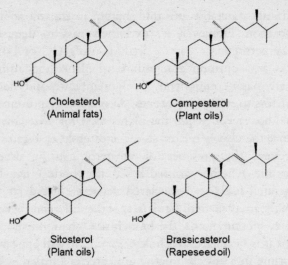

Cholesterol
(Animal fats)

Campesterol
(Plant oils)

Sitosterol
(Plant oils)

Brassicasterol
(Rapeseed oil)

Figure 6.3. Spot the difference between sterols: cholesterol from animals, sitosterol and campesterol common in many plants, and brassicasterol diagnostic of rapeseed oil.

I can't believe it's not cheese

Our fascination with cheese is summed up in Monty Python's famous Cheese Shop sketch wherein John Cleese visits Mr Wensleydale's (played by Michael Palin) artisan cheese shop. Cleese's litany of 'fermented curd' from around the world captures how we have diversified this amazing food. As Cleese says, cheese is made by curdling milk and then separating the curds from the whey. The curds can then be fermented (or not in the case of fresh cheeses like ricotta) with an appropriate bacterial culture. The resulting product helps overcome the lactose intolerance problem, since most of the lactose is left in the whey and most of the whey is removed from the curd (though many softer cheeses still contain some whey). But it also provides a food that is more transportable with a vastly longer shelf life than raw milk, while still retaining many of its highly beneficial nutritional properties. Given the diversity of cheese and the huge demand for this premium product, it is no wonder that it is on the agenda of the fraudsters.

The first cheese that John Cleese mentions is Red Leicester – characterised by its intense orange/red colour. The yellow colour of cheese *naturally* comes from the carotene pigment (see Appendix) in the grass the cows eat, which is transferred through the digestive system and mammary gland into the milk and is retained in the cheese. It was realised by English cheesemakers several hundred years ago that they could make more money if they skimmed off the cream and sold it separately or made butter from it. However, the yellow pigment was removed with the cream and the butter, so the lower fat cheese produced was white. Because the yellow colour of cheese was seen as a mark of quality, the crafty cheesemakers used yellow plant dyes to colour their cheese and pass it off for the genuine full fat cheese. The practice of colouring cheese still goes on. Indeed, since the eighteenth century Red Leicester has been coloured orange by adding annatto extract, which is made from the seeds of the achiote tree (*Bixa orellana*) and is widely used as a natural food colouring (see Appendix).

If the yellow/orange colour is a sort of visual diagnostic for cheese within our own internal databases of quality food, then you can see why producers of fake cheese or what's called by the food industry 'cheese substitute' might use yellow colouring to lure us into thinking a product is more authentic than it actually is. Food colouring can legitimately be added to cheese substitute and there is nothing illegal in producing cheese substitutes, provided the labelling clearly displays exactly what they are. Of course this is where things get messy.

If you go into your local takeaway pizza shop and buy a pizza with nice stringy melted cheese on it you would be forgiven for thinking you were buying a pizza with mozzarella cheese, as this is the most popular pizza cheese globally.

However, in the US it's estimated that only around 30 per cent of cheese on pizzas is mozzarella – either the American version or the traditional version from southern Italy. Most of our pizzas are adorned with a cheese substitute. Food scientists have done a lot of work to produce the perfect pizza cheese; real cheese is added mainly as a flavouring, while components such as vegetable oil, whey protein and salts help with the emulsification. In February 2014, the *Guardian* newspaper in Britain reported the results of an investigation, conducted by the Lancashire Trading Standards office, of pizzas being sold in their county. The report describes how officers sampled pizzas from 20 outlets and performed tests on ingredients used in hundreds of pizzas sold to customers. The Trading Standards report said 'EVERY takeaway they investigated ... has been selling customers pizzas containing "fake cheese".' The report emphasised that 'It is not illegal to use cheese analogue but it should be properly identified as such.' The analyses for other pizza toppings were little better, with four pepperoni samples containing animal species in addition to beef or pork. As many as 67 per cent (10 out of 15) ham pizzas also contained turkey DNA. Lancashire County Council Trading Standards are working with takeaway owners and wholesalers to ensure they comply with labelling regulations.

Venezuelan Beaver cheese (which incidentally has its own website, venezuelanbeavercheese.com) was among the 45 cheeses listed by John Cleese in the Cheese Shop sketch. Obviously, this is a fictitious cheese, Venezuela having no native populations of beaver and beavers being unlikely candidates for milking if they did. Yet cheese is made from the milk of quite a range of different animals, and whether owing to low yields or the general disposition of certain species when being milked, some are more expensive, based on the species alone. So how can we tell the type of animal that cheese comes from? For example, the popular mozzarella (the Italian version), manchego, feta and Roquefort cheeses are made from non-cow's milk and so methods are required to check whether these cheeses contain solely water buffalo (mozzarella), sheep (manchego, feta and Roquefort) or goat (feta) milk or whether cow's milk has also been added. Protein analysis is one of the main methods used to authenticate cheese, as different species produce different proteins, and curd, if you recall, is protein. Electrophoresis, specifically isoelectric focusing (which we discussed in Chapter 4), separates proteins based on their movement within a gel under the influence of an electrical field and has been adopted as the standard method for cheese authentication within the EU. The method can detect as little as 0.5 per cent cow's milk within sheep, goat and water buffalo milk, based upon γ_3- and γ_2-casein analyses. DNA analysis using PCR with species–specific primers has recently been shown to be a highly sensitive technique for the qualitative detection of cow's milk, even in overripe mixed cheeses (see note 6).

Concerned about the possibility of fake cheese in shops in the UK, the consumer organisation Which? teamed up with Professor Chris Elliott (author of the UK government's independent review into food crime) to investigate the possible adulteration of goat cheeses with milk from other species. Of 76 samples of goat cheese purchased from supermarkets, delis and markets around the UK, nine samples were found to be adulterated to varying degrees with sheep cheese.

The same problem has been found with other cheeses. Italian mozzarella, as we mentioned, is made from the creamy milk of the Italian Mediterranean buffalo. Whether by chance or not, the Italian word for buffalo, *bufala*, also means hoax or fraud. Mozzarella mogul Giuseppe Mandara, owner of Italy's biggest manufacturer of buffalo mozzarella, has been in the headlines off and on for years over adulteration of the cheese with common cow's milk – as well as Mafia offences and money laundering. The problem reportedly also exists in the Italian hard cheese market, where the estimate of forgeries of such brands as Parmesan,

Asiago and Pecorino Romano are estimated at 20 per cent, which is a large proportion when the value of the annual trade is in the hundreds of millions of dollars.[14] The Switzerland Cheese Marketing Association used random DNA testing to prove that about 10 per cent of supposed Emmental on grocery store shelves is actually fake, such as Italian forgeries falsely claiming to be made in Switzerland. Since the testing was introduced the fakes appear to have dwindled.[15] Thankfully, none of these frauds appear to have had health impacts. However, there are parts of the world, particularly Asia, where fraud surrounding dairy products is especially rife, with fatal and life-changing consequences for the defenceless consumers.

I can't believe it's not milk

The cow is sacred in India, but sadly the same cannot be said of the milk on offer to large numbers of the country's population. India is the world's largest milk-producing nation, with nearly 130 million tonnes of milk produced in 2012. However, the volume available per capita of the country's population was less than a third to a quarter of that consumed by most Europeans and North Americans. With demand outstripping supply, the fraudsters have devised some imaginative ways of extending the amounts available. If the levels of fraud are to be believed, then India is probably now the world's number one fake milk producer. The nature of such frauds ranges from diluting milk with water to the complete fabrication of milk.

So many reports abound in the media, including a number of documentary footages on YouTube, it is difficult to know where to start. One very detailed report in the *Bihar Times* in 2010 describes officials making a surprise visit to the main slum of Mumbai where they found people removing milk from branded packets and replacing it with water. Entire households, including children, were found with all the equipment for committing the crime: hundreds of plastic bags of branded milk, pots of questionable water, a straw to suck the milk out of the bag, a funnel to put water into the bag and candles to seal the packets afterwards. Some reports suggest that only a small amount of the milk is replaced with water, while others indicate the milk is replaced entirely with 'white water'.

The government's own investigation in 2011 by the FSSAI suggests that a huge proportion of India's milk fails to conform to health and safety standards. A Reuters report in January 2012 summarised that out of 33 Indian states, non-fat adulterants were found in all the milk samples from West Bengal, Orissa and Jharkhand. The problem appeared particularly acute in New Delhi with anything up to 70 per cent of the milk samples reported as 'tainted'. The western state of Goa and the eastern state of Puducherry reportedly conformed to the standards, with 'no indication of adulteration of their milk'.

Maneka Gandhi writing on www.mathrubhumi.com pointed out that the FSSAI offers various explanations as to why foreign chemicals were detected in so many of the milks tested. Where detergents were detected, this is explained by contamination from washed hands and failure to rinse out vessels. Formalin (or formaldehyde) is used as a preservative but its addition is illegal. Urea is reportedly added to avoid curdling during transport. The Delhi Food Safety Authority acknowledged the presence of skimmed milk powder, saying 'this is not hazardous to health, it's just reconstituted milk'. The situation is complicated by the fact that surplus milk produced at some times of the year is turned into milk powder, which is regularly mixed into fresh milk when shortages arise. Around 50 per cent of the milk sold in Delhi in the summer is reconstituted – perfectly sensible, but there are concerns that this is not declared on the packaging. There are other additions that can be logically explained if the milk is reconstituted, but there is also evidence of more extreme fakes – namely, the actual fabrication of milk.

In a perverse way, the idea of fabricating milk is not so far removed from making infant formula with mostly non-milk ingredients. But when reduced to its ingredient list and recipe, the production of this type of fake milk sounds horrific. The technology currently used to make India's fake milk is reported to have been invented by milkmen of Kurukshetra (Haryana) around the turn of the millennium. It seems that the recipe was so successful that it spread very fast, and the practice now appears to exist all over India. Commonly reported recipes for fake milk require easily accessed ingredients such as urea, caustic soda, cheap cooking oil as a substitute for milk fat, sugar, water, powdered milk, and detergents such as hair shampoo. The first step is to emulsify the cheap oil, such as soy, in water using the detergent, to produce a white frothy solution. The oil also helps to create a mixture with a smooth texture. Caustic soda is then added to neutralise the acidity and prevent the mixtures from turning sour during transport. Urea is added in place of the milk solids-not-fat (SNF) (the caseins, lactose, whey proteins and minerals). Other reported ingredients include hydrogen peroxide, formalin, glucose, ammonium sulphate and various whitening agents. The cost of preparing synthetic milk is around 5 Indian rupees per litre (£0.05 per litre; $0.04 per pint) and it is sold for up to 30 Indian rupees per litre (£0.30 per litre; $0.22 per pint). The motive is pretty clear: pure profit with scant regard for human well-being. The Delhi Health Department estimates 100,000 litres (211,338 pints) of synthetic milk and 30 tonnes of thickened milk (khoya) are being manufactured every day in the city.

The recipe for fake milk sounds more like one of the examples from Accum's *Treatise* from nineteenth-century London, but this is no crude fake. In its own way, the recipe is ingenious. The fraudsters are fully familiar with the crude testing systems that exist within the village milk cooperatives, so they have chosen the ingredients that go into the synthetic milk wisely in order to fool the test. The fat and SNF percentage is similar to real milk and blending in powdered milk disguises the taste.

There are surprisingly few reports of any adverse health impact from drinking this type of fake milk. The dilution of milk with water obviously reduces its nutritional value, while the use of contaminated water would clearly pose health risks. A study undertaken by the Indian Council of Medical Research concludes that detergents in milk cause poisoning and gastrointestinal problems. One very troubling report indicates six children dying and more than 60 falling ill after drinking adulterated milk in a state school in eastern India. The matter has clearly raised huge public concern in India, with the lack of proper monitoring appearing to be at the root of the problem. The results of the FSSAI's survey led to an affidavit being filed in response to a notice issued on Public Interest Litigation (PIL) by a group of public citizens led by Swami Achyutanand Tirth of Uttarakhand seeking to check the sale of adulterated milk and various dairy products. Notices have also been issued to Haryana, Rajasthan, Uttar Pradesh, Uttarakhand and Delhi on PIL 'alleging that synthetic milk and adulterated milk and milk products are prepared using urea, detergent, refined oil, caustic soda and white paint, which according to studies are "very hazardous" to human life and can cause serious diseases like cancer'. At the time of writing this chapter, we await developments – but we should probably not hold out too much hope for an imminent resolution.

Formula fraud

There is nothing more distressing than the sight and sound of a hungry baby in need of their feed. And there is no question that mother's breast is best. However, not all mothers are able to breastfeed their babies. The historical solution to this was wet nursing, which was a safe and effective alternative to breastfeeding that probably dates back to the very origins of our species. However, throughout the industrial period, society's negative view of the practice combined with the development of bottle feeding using artificial formulas resulted in the demise of wet nursing. An historical alternative to wet nursing would have been to substitute the mother's milk with that of other animals, but we now know that giving babies cow's milk too early can be very damaging owing to the differences in chemical composition between cow's and human milk. The earliest attempts to produce artificial replacements for mother's milk began in the mid-1800s. In 1865, using knowledge emerging about the composition of human milk, the chemist Justus von Liebig developed, patented and marketed an infant food, comprising cow's milk, wheat and malt flour, plus potassium bicarbonate. It was considered the perfect infant food. Though it was far from perfect, it was a start and in the years that followed various other formulas were developed and their compositions modified as more detailed knowledge became available about the nutritional needs of newborns and the chemistry of human milk.

Numerous commercial baby formulas now exist; however, their composition is strictly regulated. The recommendations for infant formula compositions are laid down in the Codex Alimentarius. The document is 21 pages long and includes recommendations for special infant formula compositions required for certain medical disorders. The recommended formulation is based on the premise: 'Only products that comply with the criteria laid down in the provisions of this section of this Standard would be accepted for marketing as infant formula. No product other than infant formula may be marketed or otherwise represented as suitable for satisfying by itself the nutritional requirements of normal healthy infants during the first months of life.' The standard is regularly updated as new knowledge is obtained and is currently the globally accepted standard for infant formula.

We are at our most vulnerable as newborn babies, reliant on our parents for absolutely everything, especially our first food. Whatever a mother's reasons for choosing to feed her baby formula, it has probably been a considered decision, and she is trusting that after 150 years of research there is now a replacement for mother's milk specifically designed to meet the baby's nutritional needs. It is a juncture in life when both mother and infant are exposed. And for this reason, it seems all the more incredible that people have sunk so low as to produce counterfeit baby milk that has resulted in fatalities and had life-changing health impacts.

On 29 April 2004, the *New York Times* reported that hundreds of parents in Fuyang in central China had unwittingly bought a counterfeit baby formula, in which nutritional supplements had been replaced with starch or sugar. The result was that babies manifested what local residents called 'big head disease'. The babies readily ate the formula – of course, as there was nothing else – initially displaying fat cheeks that the parents took to be a sign of good health. However, due to the low protein and nutrient content of the feed, the rest of the body failed to develop properly. The body's coping mechanism in times of deprivation is to direct resources to the brain and vital organs at the expense of other body parts. The 200 babies displaying these awful symptoms were terribly malnourished, with 13 eventually dying. The scandal was reported on Chinese state television on 19 April, and the next day Prime Minister Wen Jiabao sent a special investigation team into the region. At least 22 people who were involved in making and selling the formula were arrested. It appeared the problem extended outside the Fuyang region, with sick babies also being reported in Beijing and Guangzhou. A spot check of baby formula in stores in Guangdong province revealed that 33 per cent of brands failed to meet the national standard. Some brands tested in Fuyang had less than 1 per cent protein when the national standard is around 12 per cent.

The *New York Times* report stated that 'Investigators blamed illegal manufacturers throughout China for the problem and reported that 45 brands sold in Fuyang and elsewhere were substandard.' Sadly, it appeared that 'reports of the problem had been percolating in Fuyang for almost a year without any significant action being taken'. The most desperate aspect of this case is that the companies producing the low-cost baby food took advantage of the parents' poverty; they had no alternative but to purchase the cheaper brands because they were as low as half the price of the premium Nestlé baby formula. They never imagined of course that they were putting their babies' lives at risk. The scandal really hit home, bringing calls for greater regulation in a country renowned for producing counterfeit products. Nearly 10 years later, in March 2013, the China Food and Drug Administration (CFDA) was formed. This is a ministerial-level agency similar to the US FDA, which should streamline regulation processes for food and drug safety. Sadly, it came too late for the victims of the Fuyang baby formula fraud.

Just when you thought things really couldn't get any worse, another infant formula scandal broke. This time, the counterfeiters plumbed new depths of depravity, serving on the Chinese people a particularly inhumane and damaging food scam. The case, known as the Chinese Melamine Scandal, was a pure greed-motivated fraud perpetrated by individuals with a detailed knowledge of testing methods and seemingly no regard for the health impacts on potential consumers. The precise details of the impacts of this terrible fraud are difficult to establish with any certainty, but the Chinese have reported that the incident affected around 300,000 people, with 6 infant deaths and more than 50,000 babies hospitalised.[16]

The scandal broke on 16 July 2008, after 16 infants in Gansu province, fed infant formula milk, were diagnosed with kidney stones. Analyses of the kidney stones using HPLC and infrared spectroscopy showed they contained crystals of melamine and a melamine-related compound, cyanuric acid. There is no reason why these substances should have been present in the bodies of these infants. Melamine is an industrial chemical manufactured in huge amounts for the production of the melamine formaldehyde resins in commonly occurring plastic coatings, familiar to us in laminates, kitchenware and adhesives. Cyanuric acid, together with ammeline and ammelide, are by-products of melamine production.[17]

A WHO report confirms that on their own, none of these substances are toxic and we are normally exposed to them in very low amounts through a food's contact with packaging. Under normal circumstances, these substances are eliminated from the body rapidly with no ill effects. However, tests on animals have shown that when melamine and cyanuric acid are consumed together, as they were in the infant formula, they are highly toxic, and can be fatal. The two substances combine to form insoluble melamine cyanurate crystals, which accumulate in the kidney causing blockage and degeneration.

Sadly, this type of toxicity had been seen the year before, in 2007, in cats and dogs in the US. The outbreak was traced to the consumption of pet foods made with wheat gluten and rice protein concentrate supplied by a Chinese company. The outbreak affected thousands of pets, causing an unknown number of deaths of cats and dogs. There was no alternative but to withdraw thousands of pet products from stores. It was clear that melamine had been intentionally added to pet food ingredients to artificially increase their measured protein content. But melamine is not a protein.

This was where the fraudsters showed their expert knowledge: though melamine is not a protein, it could fool the test for protein. The quick and simple test for protein is the Kjeldahl assay, which determines the total nitrogen in a sample by treating it with sulphuric acid and measuring, with a simple titration, the amount of ammonia released.

Another test, the Dumas combustion method, works on similar principles and is equally lacking in specificity. These are routine tests used to determine the total amount of protein in foods and the fraudsters clearly knew this because neither of these tests is designed to distinguish between protein and non-protein nitrogen. So, because every molecule of melamine contains six atoms of nitrogen, adding a small amount of melamine to milk significantly increases the amount of nitrogen detected by these tests. The result is an erroneously high protein estimate. Hence, such tests made the pet food and the infant formula appear to have the necessary protein content, when in fact it wasn't protein at all.

As to where in the supply chain the melamine was added to the infant formula is not at all clear. It has been suggested that the melamine used in the fraud may not have been the highest grade product since this would have been expensive and free from the cyanuric acid. A cheaper impure by-product may have been used, which would account for the renal toxicity seen in both the humans and the animals. No matter what the mechanism for the introduction of the melamine into the baby formula, investigations showed that the fraud was widespread and ultimately extremely damaging, not only to the vulnerable victims, but to the entire Chinese dairy industry from top to bottom. Inevitably, the real victims, the less wealthy Chinese parents of newborns, were left wondering what food they could trust to give their babies. Ironically wet nursing was reportedly on the increase.

The BBC provides a timeline[18] of the Chinese melamine scandal – the first deaths and arrests, the increasing tally of sick babies, discoveries of more melamine-laden products (pet food and eggs), nationwide alerts and bans on Chinese-made food products, bankruptcy and, finally, death sentences. It reveals the full horrors and impacts of this terrible episode – let's hope the world never sees the likes of this again.

Such frauds, which are thankfully rare, have profound effects on any food industry at all levels in the supply chain, including innocent farmers. The producers that use the adulterated foodstuff suffer economic losses, as do the outlets for named products or brands, owing to the loss of public confidence, which can take years to recover. The obvious reaction to such a scandal is to demand increased food testing. However, the melamine scandal highlights the impossible position that the protection agencies are placed in when they need to test for an unknown or totally unexpected criminal activity. Whether it is an as yet unrecognised adulterant chemical or adulteration practice – analytical science simply doesn't work this way. As we saw from Chapter 2, analysts need to have some idea of what they're looking for and pose the questions accordingly.

Comprehensive testing to ensure ultimate food quality and safety is the unrealistic 'Holy Grail' of food fraud forensics. As we have mentioned before, foods are chemically highly complex and we simply can't analyse for everything, especially for substances that have yet to be recognised as threats. The situation is quite different once a particular fraudulent practice or adulterant chemical has been identified – then testing can be harmonised internationally, with the most appropriate analytical methods being recommended and purity criteria defined. A horizon-scanning approach aimed at predicting the next big food fraud is on the agenda of the protection agencies but this type of preemptive approach will never be fully effective – the food cheats are working equally hard on their side.

Indeed, who would ever have predicted that hydrolysed protein reportedly made from ... wait for it ... *scrap leather* would be added to milk! This was reported in 2011 by the Hong Kong government's Centre for Food Safety (CFS), which had been conducting regular testing on milk and milk products. Apart from this being bizarre in the extreme, the practice could be highly damaging to health owing to metallic contaminants used in the manufacture of leather, which could be transferred with the hydrolysed protein. The CFS is monitoring milk extensively to ensure that any sales in Hong Kong comply with the legal standards and are fit for human consumption. The tests for this abhorrent activity involve the use of ion exchange chromatography to detect chromium (derived from the tanning agent) and hydroxyproline (a characteristic leather protein amino acid).[19]

The more we have learned about food adulteration, the more it seems like the worst sort of practical joke – taking advantage of trusting people. Yet, as with practical jokers, the acts expose more about the criminal's own character deficiencies than those of their supposedly gullible victims. What is especially dispiriting about the food cheats' deceptions is that their trickery can affect unsuspecting consumers in life-changing ways.

CHAPTER SEVEN

Seasoned Criminals

Spices, more than any other food commodity, have helped shape our modern world. In the fifteenth century, Europe awoke to the spices of Africa and Asia. These spices motivated nations to send explorers across vast oceans in search of new sources and new trade routes. By the sixteenth century countries were competing for control over those resources and routes. It catalysed both the making and the breaking of cultures.

In 1667, the Dutch negotiated a trade with the British: the island of Manhattan in exchange for an Indonesian island 1/30th the size – Run (or Rhun). While Manhattan, located at the mouth of the Hudson River, lay at the door to the fur trade, Run had an even more lucrative resource: nutmeg.

If it weren't for the small evergreen tree *Myristica fragrans*, the 3km² (about 741-acre) island of Run, and the other nine volcanic Banda Islands, would probably have been ignored by Europeans. The species is thought to have evolved on these and other islands in the Moluccas, and at the time was found only there. The trees produce up to 20,000 small yellow pear-shaped fruits each season. Though the fruit is edible, it is the pip at its heart that is of real interest – this is nutmeg. Woven around the dark nutmeg seed, in a manner reminiscent of a 1970s macramé plant hanger, is a red fleshy covering known as the aril. It's a plant's enticement to animals to eat the seed and help it disperse its progeny. Another well-known aril is the edible flesh of the lychee. The aril surrounding nutmeg is ground into another spice – mace. Two expensive spices with culinary value, hallucinogenic effects and medicinal uses, from a single rare plant found only on a small group of remote Indonesian islands – no wonder battles were fought.

After the British left Banda, the Dutch managed to maintain a nutmeg monopoly for 100 years. It was a French horticulturalist who would eventually bring it to an end. Pierre Poivre (a name so perfect for a chapter on spices that one might think we made it up) managed to smuggle out some nutmeg trees, which he brought back to Mauritius (then the Isle de France) where he was administrator. He planted them in the botanical garden he was building and, with a few flicks of a spade, ended the Dutch nutmeg monopoly.

While nutmeg may have been one of the rarest spices, it was the price of pepper that reflected the state of the economy for centuries. Black pepper is the dried fruit of the *Piper nigrum* vines of southern India. Pepper made its way out of India thousands of years ago, however, as evidenced by its presence in the nostrils of a 3,200-year-old Egyptian mummy. Pepper became the most widespread seasoning of the Roman Empire; it was transported via secure trade routes from India, up the Red Sea and the Nile River to Alexandria, where it was then shipped to Italy and Rome. Italy maintained a monopoly on the black pepper trade during the Middle Ages, from the fifth to the fifteenth century.

Pepper was of such value that it was often used as currency or collateral, and was sometimes referred to as black gold. Hints of its value linger even in today's Dutch language – *peperduur*, which translates as 'pepper expensive', is an expression for something that is very costly. After the Italian reign over the pepper trade, the Portuguese got in on the action. With the signing of the Treaty of Tordesillas in 1494, Portugal gained exclusive rights to half of the black pepper-producing land. Yet smugglers continued to move it through the old trade routes via Alexandria and Italy. The Dutch and English moved in and by the end of the fifteenth century black pepper was relatively common in Europe and its price began to decrease. Today, pepper remains a very large portion of the international spice trade. In 2012, the value of the pepper market was estimated at £1.08 billion (US$1.77 billion), with Vietnam exporting around 45 per cent of the world's pepper. And so it seems fitting that we should begin our sordid story of spice-swindling with pepper.

Pepper is not all it's cracked up to be

The UK Guild of Pepperers – *Gilda Piperarorium* – dates back to the twelfth century (circa 1180). Pepper merchants traded in precious black pepper, but also dealt in sugar, dried fruits and alum. This was perhaps the most important of the food guilds as it was the pepperers who were granted the responsibility of holding the king's weights and measures.

In the fourteenth century the pepperers and spicers joined forces and became the Worshipful Company of Grocers. It was then that this livery company was granted the office of garbling. The word originates from the Arabic *garbala*, which means to sift, and this is exactly what they did. Using a set of sieves, the Grocers' Company would check and sift all the spices that landed at the London docks looking for impurities such as gravel, leaves and twigs. They used their knowledge of spices to determine the purity of the product. Every spice was checked by a garbler before it could be sold; it was a compulsory and costly venture and the Grocers' Company had a monopoly on it.

And so life continued for several centuries until a series of events in the seventeenth century triggered the decline of power for the grocer-garblers. The East India Company sought exemption from garbling, and though it wasn't granted, it was the first sign that power was shifting from the guilds to big business. Then the pharmacists, who until then had been part of the Grocers' Company, separated into the Worshipful Society of Apothecaries in 1617. Drugs and oils were being rampantly adulterated and detection of such fraud required chemists rather than garblers, hence their decision to branch off. With the decline of the guilds' power, new opportunities opened up for the adulteration of spices. By the nineteenth century it was the grocers and merchants themselves who had become the swindlers, earning a reputation for corruption and fraud.

The adulteration of pepper was well known in the early nineteenth century, but chemist Frederick Accum quantified the extent of it. Using only a glass of water, Accum discovered that approximately 16 per cent of pepper sold in London at the time was counterfeit. The artificial peppercorns were made by combining the residue of linseed (after it had been pressed for its oil), common clay and a portion of cayenne pepper. This concoction was then pressed through a sieve and rolled in a cask to achieve the appropriate peppercorn size, shape and appearance. Accum simply tossed the peppercorns into a bowl of water and those that were not pure would disintegrate in the water. In Accum's time, anyone discovered making or selling adulterated pepper was fined £100 (US$157), which is the modern-day equivalent of about £10,000 (US$15,700).

Less effort was required to adulterate ground pepper. Small amounts of genuinely ground pepper were bulked out with the floor sweepings from the pepper warehouses, known as pepper dust, along with some cayenne. Sometimes an even more inferior product to pepper dust was added, which was known as the 'dirt of pepper dust'. This was no doubt more dirt and dust than pepper.

It was the physician Arthur Hill Hassall (we first met him in Chapter 2), wielding his microscope, who opened Londoners' eyes to all that could be added to ground pepper. Renowned first for revealing the creatures inhabiting London's water supply in 1850, Hassall turned his lens on food and was dismayed at what he found. It was Hassall who first identified the mite in brown sugar that caused grocer's itch. The microscope was a new tool in the forensics toolkit.

Hassall studied the genuine form of pepper, cell layer by cell layer. With a keen understanding of what the real product looked like, he quickly set to work identifying the fragments that were of non-pepper origin in the samples of ground pepper that he collected. In all, he identified linseed, mustard seed, wheat flour, pea flour and ground rice amid the pepper fragments.

These days dried papaya seed is the most common adulterant in peppercorns. They are cheap, only a fraction smaller than the real thing and easily available. Just as with Accum's detection of counterfeit peppercorns, only a glass of alcohol is required to tell the papaya seeds apart, because the papaya seeds float while true peppercorns sink. This isn't a foolproof method, however, as immature pepper seeds can also float.

Another long-used method of adulterating peppercorns is to coat them with paraffin oil and burnt diesel oil. The coating adds weight to the peppercorns, helps preserve them from fungal infection and also provides them with a black polished shine. In 2014, 18 tonnes of pepper coated in these potentially carcinogenic oils was seized in a single raid of a warehouse in Chennai, India. It was worth about £105,000 (US$165,000). Consumers should be able to identify this pepper easily as it has a distinct oil or kerosene smell and is darker and more polished than the real thing. Yet the Consumer Association of India stated in response to the raid that consumers don't take the time to check their spices. If people are failing to catch adulteration that is visibly detectable in a whole spice, what hope is there of finding it once these spices are ground?

Papaya seeds make their way into ground pepper too, but it's difficult to know whether they were added by the grinders or the wholesalers of the peppercorns. Millet and buckwheat flour have also been used to add weight to ground pepper. Though the blatant addition of dirt of pepper dust from Accum's day may be ended, there are still guidelines as to how much filth can be included in pepper before it is considered an adulterant. The FDA states that pepper can contain no more than 1 per cent of

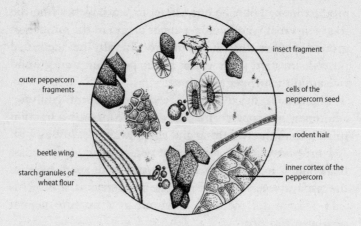

Figure 7.1. The fragments in a ground pepper sample that Hassall may have spied under his microscope, complete with rodent hairs, beetle bits and wheat flour adulterant.

insect-infested and/or mouldy pieces by weight. Nor can it contain more than an average of one milligram (0.03oz) of mammalian excreta per half kilo (per pound). You may also be relieved to learn that the FDA recommends legal action if the average number of insect fragments per 50 grams (1.7oz) of pepper exceeds 474, or if the average number of rodent hairs per 50 grams (1.7oz) exceeds one. Given the nature of spice harvest and storage, it isn't terribly surprising that a few insects and rodent hairs find their way into the product. But this is surely a case of accidental contamination – poor hygiene practices at worst. It is the intentional adulterants that are far more worrisome, no matter how unappealing mammal excrement in your pepper may seem.

Spent spices, Sudan Red and other spice swindles
Pepper is not by any means alone in its susceptibility to fraud. Indeed, adulterating spices is a relatively straightforward process, particularly when you compare it

with making a fake egg, mixing bogus milk or fashioning false mutton. It is a ground product in which foreign bodies can easily be masked. There are economic incentives, as spices are relatively expensive, and they are grown in some of the most impoverished countries in the world by subsistence farmers. There are many opportunities for substitutions between the farmer and the spice rack. The scales are definitely in favour of the fraudster.

What makes consumers more vulnerable to fraud is that many of us are reasonably ignorant of the properties of these exotic plant parts – how they should look and smell and behave as they infuse into our cooking. Such ignorance may have us choosing the paprika that has had its colour bolstered with carcinogenic dyes over the pure product. Our perceptions of purity can even lead us astray with plants more familiar to most of us. Dark green dried oregano, for example, is more visually appealing to consumers. Yet this is often an indication that dried leaves from other plant species, such as *Cistus incanus* (rockroses grown in the Mediterranean region), have been added to the mixture. The adulterated product is more tempting than the more pallid, yet genuine, product.

The list of adulterants in spices is long. We have highlighted the most interesting ones below, but should you wish to see a more extensive list, we refer you to the USP Food Fraud Database, which is searchable online.

Cayenne

This is a powdered mix of the ground seeds and peppers from different *Capsicum* species, but mostly *Capsicum annuum*. This spice has been adulterated with ground rice, mustard seed husk, sawdust, brick dust, salt and turmeric. To cover the addition of these adulterants, the mixture is coloured with red lead (lead (II, IV) oxide), which is used in manufacturing lead batteries and rust-proof primer paints. Chronic exposure to red lead may cause lead poisoning as well as other health implications.

Chillies and chilli powder

In both whole and powdered form, chillies have had adulterants added to increase their weight or make them appear to be a superior product. Brick powder, sand and dirt have all been used to add weight to chillies. Malachite green – the carcinogenic antifungal that we mentioned associated with fish farms and will mention again below in relation to coriander – has been used to make green chillies appear more vibrant. Sudan dyes, which are red dyes used to colour oils, waxes, shoe and floor polishes, have been used in chilli powder. They are not legal for use in food in the EU owing to their potential carcinogenic properties, allergenic activity and mutation-inducing effects. In 2005, more than 350 food products were removed from UK shelves because Worcestershire sauce used in their production contained chilli powder that had been adulterated with Sudan I.

Cinnamon

This spice is obtained from the inner bark of the tree *Cinnamomum verum*. The inner bark from other species within the same genus, namely *Cinnamomum cassia*, is referred to as cassia or Chinese cinnamon and is often sold as cinnamon. Sri Lanka is the world's biggest producer of cinnamon. The powdered version of this aromatic wonder has been adulterated with coffee husks, sago, wheat flour, potato flour and arrowroot powder. Cinnamon has also had the essential oils that impart its discernable smell and taste removed and then it has been re-dried and sold deceivingly as the intact product. Hassall was the first to be able to detect such tampering as the starch granules in the cinnamon become distorted and irregular when it has been boiled in order to remove these oils.

Coriander

This is the ground seeds from the herb of the same name, although the leaves are referred to as cilantro in some countries. Ground coriander has had rice husk, wood fibre and salt added to increase its weight. The most frequently reported adulterant, however, is cow dung. There are two versions of cow dung – one being organic and produced from the business end of a cow, and the other being synthetic with significant potential to harm human health. Dried cow dung, shaped into cow cakes, is a popular biofuel in India, but it can also be ground up and used as a fertiliser, cleaner, polisher, tooth polish and skin tonic. It even plays an important role in the preparation of Ayurvedic medicines and deity worship. In fact, its many uses have led to a supply shortage and this has triggered the development of synthetic dung powders, namely Auramine-O and malachite green. Both of these chemicals are toxic, particularly Auramine-O, which causes multiple organ failure if inhaled or ingested. Despite being illegal to sell, these chemicals remain relatively accessible and Auramine-O has emerged as a common drug of choice for suicide in India. There appears to be no distinguishing the genuine product from the synthetic – both are referred to as cow dung powder, including when they are named as an adulterant in spices. While a little dried cow dung powder in your coriander may seem somewhat unpalatable, it is a far better option than its synthetic alternative.

Cumin

This foundation for any good curry is the seed of *Cuminum cyminum*, which belongs to the same family of plants as caraway, celery, carrots, parsley and parsnips. India is the world's biggest producer and consumer of cumin. It isn't known as an expensive spice, but low yields inflated prices in early 2015; hot weather delayed planting and farmers shifted to other crops after prices dropped in 2014. Probably due to this short supply and higher value, cumin was at the centre of a food fraud scandal far more dangerous than Horsegate.

In late 2014 and early 2015, a number of ground cumin products in the US and Canada were recalled as undeclared peanut protein was detected in the mix. In response, the UK started testing cumin products and the FSA found undeclared almond protein. Hundreds of products containing the spice – from fajita mixes to falafels – were recalled as a result of the adulteration. Peanut shells and almond husks cost nothing, and when ground up they can be mixed in with ground cumin to bulk it up. Yet this creative cost-saving strategy can be lethal to those with nut allergies as the husks and shells often have remnants of nut attached. US resident Jillian Neal is deathly allergic to peanut products. Luckily a fast response from her family and medical professionals in November 2014 saved her after she ate chilli made with a mix containing the contaminated cumin. While warnings were issued advising allergy sufferers to avoid cumin products, the spice is often not declared on the label as it is used in such small quantities and can be considered a trade secret. But it is not just the ground version of this product that has been involved in fraud. Grass seed coloured with charcoal dust has been sold as whole cumin. Needless to say, if the cumin leaves your hands black after you touch it, you've been duped.

Ginger

The rhizomes of *Zinziber officinale* are used fresh, candied, dried or powdered. When Hassall looked at the ground product under his microscope in the nineteenth century he found two-thirds of the samples he examined were adulterated with one or more of the following products: wheat flour, cayenne pepper, potato starch, sago, turmeric, mustard husk and ground rice. And these adulterants formed the majority of the product.

Nutmeg

There have been no reported cases of people trying to substitute some other nut or seed for whole nutmeg; it seems unlikely people could be so easily fooled. However, much like with cinnamon, there were historical cases of the volatile oils of the nutmeg being extracted through distillation before selling the whole nutmegs. These nutmegs would feel light, dry and brittle in comparison with an unaltered nutmeg. Ground nutmeg has had coffee husks added to increase its weight.

Saffron

Known as the most expensive spice in the world, and rightly so, saffron is the stigmas from the domesticated flower *Crocus sativus*. The flowers are harvested by hand and it takes nearly 30 hours of labour to harvest the 100,000 stigmas needed to make up a kilogram of saffron, which is worth an average of £6,500 (nearly US$10,000). The world's largest producer of saffron is Iran. Saffron threads can be bulked up by including wasted bits from the saffron flowers and weight can be added by infusing the saffron with syrups, glycerine, oils, gypsum, borax, starch and other distasteful products. There is an extensive list of items that have been dyed to mimic the vibrant crimson stigmas of the real thing, including parts of other flowers, such as marigolds, carnations, poppies and safflower. Somewhat more daringly, people have disguised beet fibre, capsicum, corn silk, grass, onion and silk fibres as saffron; they have even fashioned threads from gelatin and the fibres of dried animal meats.

The list of dyes used include quinoline yellow, which is a permitted food additive in Europe (E104) and Australia, but is not permitted in food in Canada and the US; ponceau 4R, a synthetic colourant that has been approved as a food colouring in Europe (E124), Asia and Australia, but not in the US or Canada; tartrazine (E102); and sunset yellow (E110). All of these colourants have been linked to hyperactivity in children. More alarmingly, the following dyes, which are considered toxic and/or carcinogenic, have been used to colour fake saffron: methyl orange, which has mutagenic properties sufficient to warrant avoiding direct contact with the substance; naphthol yellow; and red 2G (E128), which Europe banned owing to health concerns in 2007 and which has also been banned in Australia, Canada, Israel, Japan, Malaysia, Norway and the US.

Salt

Salt has reportedly been adulterated with white powdered stone and chalk. Chalk can easily be identified by stirring the suspicious salt into a glass of water. If chalk is present, it will turn the water white. In 2012, Polish health authorities ordered the recall of 230,000kg (500,000lb) of food – mainly pickles, sauerkraut and bread – that was suspected of containing industrial road salt rather than table salt. Agencies tested the salt intended for de-icing roads for dioxins and heavy metals and stated that it was not harmful to human health, but they recalled the food anyway as a precaution. Several years earlier, China, the world's largest salt producer, was mixed up in salt scandals when over 700 tonnes of industrial salt was found on the market. The fraudsters were packaging up industrial salt into small authentic-looking packages for the food market. Industrial salt doesn't contain iodine as edible salt does. And the heavy metals and dioxins in the industrial version can affect mental and physical development and potentially impair reproductive function. In Shanghai, food cooked using the salt resulted in the death of a 38-year-old man and the hospitalisation of 25 others. The symptoms include nausea, headache, vomiting and a rapid heartbeat. The fraudsters can make about £0.07/kg (US$0.26/lb) on the scam.

Turmeric

The deep yellow turmeric powder that many of us are familiar with comes from the plant *Curcuma longa*, which comes from the same family as ginger. The rhizomes, which look very similar to fresh ginger, are boiled and then dried before being ground into powder for a number of uses, including cooking. This spice has a long list of adulterants associated with it, including starch, sawdust, rice flour, yellow clay and chalk powder. As with saffron and chilli, it has had a number of unpleasant dyes added to it to mask the addition of these cheap fillers. Sudan dye has been used, as well as lead chromate, which is used to colour paint and can lead to death if inhaled or swallowed. Metanil yellow is the most frequently added dye found in turmeric and, in fact, is the most common illegal food colourant encountered in India. Studies have shown that prolonged consumption of metanil yellow can have long-term health implications, including changes in the nervous system, lung function and organs as well as reductions in fertility. Turmeric is reasonably unique to the spices, however, in that it has a history of being adulterated, but is also an adulterant. It has been added to cayenne pepper, saffron and paprika to enhance their colour.

Vanilla

Vanilla beans are the fruit of orchids of the *Vanilla* genus, but commercial vanilla is generally derived from the species *V. planifolia*. The vine-like plant is found in tropical and sub-tropical regions globally, but is thought to have originated in Mexico and Central America. Vanillin is the chemical compound that is largely responsible for vanilla's characteristic flavour and smell, but there are in fact about 200 aromatic compounds altogether that form the complex flavour of natural vanilla. Real vanilla extract is made by soaking vanilla beans in a solution containing 35 per cent ethyl alcohol in water. It is the second most expensive spice on the market and, therefore, a target for adulteration.

Tonka beans have been used as a vanilla substitute as they contain the chemical compound coumarin, which has a vanilla-like odour and taste. However, it is also moderately toxic, particularly to the liver and kidneys, and has been banned by a number of countries as a food additive. Oddly enough, the compound isn't found in real vanilla, so it is a good target for detecting adulteration.

There is also artificial vanilla on the market, which is a much cheaper option. Most artificial vanilla is a solution of straight vanillin. It lacks the other 199 compounds in real vanilla that provide the complex flavouring, which makes it a somewhat inferior product. Most vanillin is synthesised as demand for it within the food and beverage industry as a flavouring far outweighs what could be produced by the beans alone. It used to be made from lignin – a main constituent in the cell walls of plants, which is a by-product of the pulp and paper industry. Vanillin is one of the molecules that is infused into alcohol that's aged in oak casks, as the lignin in the oak breaks down. Most vanillin used to be produced from the waste product from pulp mills – a good use of otherwise worthless material. However, there were environmental concerns about this production method as it required using highly corrosive strong bases that then had to be neutralised later with strong acids – all with environmental implications. Most vanillin today is synthesised from the petrochemical precursor guaiacol, which is a naturally occurring compound that also contributes to the flavour of roasted coffee. It is synthesised for industrial use by methylation of catechol, a common building block for organic compounds. This is a more environmentally friendly method of synthesis, but it is also more expensive. Alternative methods are being explored, including a method developed by scientists at the Universiti Putra Malaysia that converts the lignin in sawdust without the need for harmful chemicals. In 2007, Japanese scientist Mayu Yamamoto won the Ig Nobel Prize for chemistry for extracting lignin from cow dung and converting it to vanillin.

Of course, in terms of food fraud, all of this artificial vanilla can be used to adulterate or replace real vanilla and sell for 200 times the price. By law, if vanilla contains no more than one ounce of synthetic vanillin per unit of vanilla extract, it must be labelled as 'vanilla flavoured' or 'vanilla-vanillin flavoured'. More than that, and it has to be labelled as 'artificial vanilla'. It is relatively easy to detect whether artificial vanilla has been added to pure vanilla to eke it out as it contains ethyl vanillin, piperonal and sometimes coumarin, compounds not found in pure vanilla.

All of these spices can be adulterated with spent spices – those that have passed their potency period. We all have examples of these in our cupboards, such as a kilo of turmeric from a bulk goods store that was just too good a deal to pass up. It might still look bright and yellow, but the essential oils and volatile organic compounds that gave it the aroma and flavour we so desire have long since dispersed into the atmosphere.

Spice blends, such as garam masala and curry powder, are equally prone to all of these adulterations, with the added complication that it is unclear at what stage in the processing the fraud has been committed – it could have been in the blending process, or in the actual spices that were blended.

Finding the fakes

Unlike some of the food frauds described in previous chapters, the adulterants found in spices are so diverse that an entire tool chest of analytical methods is required to reveal the fraud. While cheap oil is substituted for expensive oil or one species of fish is labelled as another, any number of things, organic or synthetic, can be ground up and mixed into a spice. To make matters even more complicated, spices are very varied in their botanical origins – they are seeds (cumin and coriander), berries (pepper), rhizomes (ginger and turmeric), roots (horseradish), bark (cinnamon), floral parts (saffron) and even floral buds (cloves). Tests that might work for a seed spice may not be appropriate for a root spice. This makes spices a particularly challenging commodity for analysis.

Accum devised some of the first chemical tests to look for adulterants in spices. The detection of red lead in cayenne, for example, could be done by shaking a sample of the spice in a sealed vial with hydrogen sulphide water; if lead was present, it would turn to a muddy black colour (though I'm not sure how many households had access to hydrogen sulphide water). Yet there are still any number of quick tests to look for adulterants that can be done with little more than what can be found in the kitchen cupboard, and wet chemistry methods (generally mixing liquids together in glass beakers) are still used by analytical labs for authenticating spices. There have been numerous efforts in India, on behalf of government agencies and media outlets, to publicise such tests. Here are some examples:

- Metanil yellow is conveniently an acid–base indicator – it changes colour within a certain pH range. Therefore, to test whether it is present in turmeric, add just a few drops of acid, such as citric acid (lemon juice) or hydrochloric acid (found in many drain cleaners). If metanil yellow is present, there should be a colour change from yellow to red that persists. Acetic acid (vinegar) isn't a strong enough acid to cause the colour shift.
- To determine whether starch has been added to a spice, add a few drops of iodine solution. When iodine comes in contact with the amylose component of starch, a deep blue/black colour forms. If there's no starch, the iodine will remain orange or yellow. This test won't work for ginger or turmeric as they are both roots that naturally contain starch.
- Suspicious saffron threads can be soaked in tepid water. Both the genuine product and the dyed fakes will release a yellow colour into the water.

The genuine product will continue to slowly release a pure yellow colour while the threads retain their vibrant colour. The fake threads, however, will start to lose their colour and the water will become orange with the intense colouring.

- To test for cow dung in ground coriander, sprinkle it into a glass of water. Dung will apparently float, but more importantly, once it is wetted it should smell as you would expect (not like coriander).

Though most of us don't have easy access to a microscope, and nor would we know what to look for if we did, it is still a powerful tool in the detection of adulterants. The American Spice Trade Association (ASTA) recommends using microscopic analysis to detect grains, hulls, starch, non-declared herbs, floral waste, buckwheat, millet seed and coffee husks in a number of spices. No doubt it is also the best method of ensuring that the number of rat hairs and insect fragments are within the legal limits too.

Chromatography (commonly linked with mass spectrometry) offers a set of particularly useful tools for detecting spice adulteration. As we have mentioned previously, chromatography is good for pulling apart (and sometimes quantifying) components in a mixture, even if they are extremely similar to one another. HPLC, an analytical technique we've mentioned in several previous chapters, is widely used in spice authentication; once again, it is looking for chemical fingerprints of either the pure spice or the adulterant. For example, in the case of saffron, HPLC can be used to detect the three main compounds of the genuine saffron crocus: crocetin esters, which give saffron its yellow colour; picrocrocin, which gives saffron its unique flavour; and safranal, which is responsible for saffron's aroma. Together, these three compounds determine the quality of saffron. If one of these compounds is missing, particularly picrocrocin, which is unique to the *Crocus* genus, then it is unlikely to be true saffron and further testing might be required. The analyst may also be looking for a specific marker of a known adulterant – metanil yellow or Sudan dyes in turmeric, chilli or curry powders, for example, will show up as an unexpected spot on the chromatogram.

Gas chromatography (GC), which uses a gas as the mobile phase, is particularly useful for analysing easily vaporised compounds, such as the volatile organic compounds that are responsible for many of the aromas and flavours associated with spices. It can therefore be used to detect whether spices are spent – that is, their essential oils removed. The GC profile of a spice will change over time as the volatile organic compounds dissipate. GC can confirm the presence of papaya seed in pepper as papaya contains the compound benzyl glucosinolate, while pepper does not.[1]

GC/MS is also used to detect pesticide residues in spices. This is a topic worthy of further discussion, which we will do in Chapter 9 when we discuss fruit and vegetables in more detail, particularly those labelled as organic when they are clearly not. In terms of spices, though, testing in the Pesticide Residue Research and Analysis Laboratory at Kerala Agricultural University in India in 2014 revealed that chilli powder, cardamom and cumin can all be highly contaminated with pesticide residues that are beyond the legal limits. In 50 samples of cardamom tested in 2011, 74 per cent of the samples had pesticide residues, including DDT.[2] These pesticides are in the soils where the crops are grown. So, while this contamination is not intentional, it can still be considered a form of fraud as consumers are led to believe that the products they are sold have met certain safety standards, which is obviously not the case. We promise to return to this.

Spectroscopic techniques are another set of tools for authenticating spices. The basic principle behind this group of techniques is that light (usually in the near- or mid-infrared spectrum) is shone on or through the sample being analysed and the light that is reflected back is measured. Any wavelengths of light that are not reflected back or do not pass through the sample are absorbed by the sample. The chemical bonds in the sample get excited by the light radiation and will vibrate and absorb light at different wavelengths depending on the type of bond; a carbon-hydrogen bond will dance to a different radiation tune from a carbon-carbon bond. Also, carbon-hydrogen bonds in an aromatic ring, such as that found in the principal component of cinnamon, cinnamaldehyde, will absorb a different wavelength of light from the carbon-hydrogen bonds outside the ring. In other words, much like people on a nightclub dance floor, the carbon-hydrogen bonds behave differently depending on who's closest to them.

The output of the analysis resembles the profile of a cave ceiling, with stalactites of varying lengths and widths. The shape, magnitude and absorption band of each stalactite-like peak holds a piece of information about the molecular structure of the compound being analysed. It's a near-instantaneous molecular signature. Of course, there's lots of noise in these data as well, so statistics, known as chemometrics, are used to sort out the irrelevant information from the useful information and to tease apart more complex data. These types of methods have proved useful in quantifying the amount of buckwheat or millet in ground black pepper as well as the adulteration of turmeric with chalk powder.

We could continue to list the various analytical techniques, but we fear that if we start throwing around terms like near-infrared hyperspectral imaging you may start to glaze over. The point is that all of these analytical methods are used to separate the compound under investigation into its various constituents. Some of these techniques have advantages over others depending on what is being investigated and the constituent that is being sought. The advantages of these methods in general are that they are fast and cheap to run and the equipment required is standard for most analytical labs.

The disadvantage is that the molecular structures of some substances can be indistinguishable. Pure cinnamon and cinnamon adulterated with corn starch, for example, give nearly identical spectra and it is only after secondary processing of the data that very small differences can be seen. Furthermore, it has been shown that the molecular signatures of spices change over time. Hence the term 'signature' rather than 'fingerprint'.

DNA-based methods are also used in spice authentication, particularly for identifying plant-based adulterants. However, extraction of DNA from plants is somewhat less straightforward than from animals. The protocols used to pull the genomic material from the sample not only depend on the plant being analysed but also on what part of the plant is used – seed, root, flower and so on. The methods that we have described in earlier chapters, such as barcoding and PCR, have been used to screen saffron products for non-*Crocus sativus* species, to separate cinnamon from cassia and other *Curcuma* species from turmeric, and to identify papaya in pepper. The utility of DNA-based methods will continue to increase as more primers are developed to detect the common adulterants of certain spices – dried red beet pulp in chilli powder, for example. The precision of DNA-based methods, when they work, make them appealing in terms of gathering indisputable evidence about the presence of life-threatening allergens such as nut protein in cumin and other cases of fraud.

The nutty spice incidents were a wake-up call reminding us that spices are arguably the most fraud-vulnerable commodity in our food supply. Vegetarians and vegans who only glanced at the chapters on meat and fish are as subject as omnivores to these fraudulent flavourings. No religion and no culture is immune, though some are obviously more exposed. Spices are in everything. More concerning is that they may not be listed on the label.

In January 2014, a 38-year-old English man with a serious nut allergy died after eating a takeaway curry.

The owner of the curry restaurant has been charged with manslaughter by gross negligence and is standing trial as we complete this chapter. As a result, the details are not yet available, but one has to wonder whether adulteration of the spices used in the curry will be a main argument from the defence? With nut protein in cumin, it is really rather a miracle that there have not been more cases of people reacting to their takeaways.

Not only do food forensics have to try and resolve the crime, they also have to be robust enough to stand up to the trial process. Bart Ingredients is one of the spice suppliers that recalled some of their products during the nutty spice fiasco. At the time of writing this chapter, they were raising doubts and questioning the accuracy of the tests being used by the FSA to identify the nut protein. The current methodology is to use ELISA, which you may recall from the meat chapter. The test depends on an antibody to almond protein recognising and binding to any almond protein found in the sample. However, Bart Ingredients has argued that the test can give false positives for almond protein as the antibody can also bind to proteins from the spice mahaleb, which, interestingly, tastes of bitter almonds. Mahaleb comes from the tree *Prunus mahaleb* – a species of cherry that's cultivated for the spicy seed at the core of the cherry stone. In small amounts, it is therefore conceivable that there has been some cross-contamination in the growing, processing or storage and transport of these two spices. If found in large amounts, however, one would have to suspect its presence is more intentional.

As of December 2014, new EU legislation was introduced that requires restaurants and takeaways in Europe to inform customers if their food may contain any of the 14 most common food allergens: celery, gluten-containing cereals, crustaceans, eggs, fish, lupin, milk, molluscs, mustard, nuts, peanuts, sesame seeds, soya and sulphur dioxide. This is no simple task if you consider that wheat, nuts, peanuts, mustard and soy have all been found as adulterants of commonly used spices! While the restaurant shouldn't be blamed because the paprika they used was corrupted well before it came into their hands, it is unfortunately their premises that will be wrapped up in the scandal. It is on their floor that the anaphylactic reaction will occur, their diners that will be affected by the scene and it will be their name in the headlines. It is a thought that must have restaurant owners taking a thorough look at their supply chains.

The European Spice Association, Seasoning and Spice Association and ASTA – all groups that represent spice companies – have developed papers on spice adulteration in an attempt to educate their members on the issue. The task of combating fraud once spices are on the shelves and in other products seems an insurmountable task. It therefore seems as though the industry itself will have to be a leader on this front. A return to the Pepperers' Guild, if you will, where there is pride (and a premium) in providing a genuine product.

Of all the types of food, we suspect it is spices that are most likely to be embroiled in food fraud scandals in the future. For the last five years, there has been a steady increase in the volume of EU spice imports, growing an average of 4.1 per cent per year. The value of these imports, on the other hand, has skyrocketed with an average price increase of 8.3 per cent per year.

The obvious economic incentives aside, spices are also most prone to the effects of global climate change. The spice trade is largely dependent on the production from developing countries; they provide 57 per cent of total EU imports. These countries are at the forefront of climate change and are already experiencing changing rainfall patterns. These changes are resulting in severe and prolonged droughts in some regions and frequent brutal flooding in others. The eastern Himalayas have seen a steady decline in one of their biggest cash crops – cardamom.

Warmer and drier winters have allowed a fungal blight to flourish, killing crops and forcing farmers to turn to less valuable crops.[3] Low yields of certain crops will need to raise the fraud alert flags.

Climate change may also provide new opportunities to grow spices where they have not been grown traditionally. The UK, with its insatiable appetite for curry, has already started to consider that land that is currently marginal might become ideal for growing some of the spices currently imported. Capsicum and chilli pepper production has steadily increased in the US as well as in parts of the EU. However, there needs to be sufficient agricultural land or enough financial incentive to switch traditional crops over to these new and potentially risky ventures. There needs to be sufficient demand and plenty of cheap labour available, as many of these crops are labour-intensive, particularly at harvest. The official forecast from the industry is that the demand for spices will continue to outpace the supply.

Spices also present a challenge to consumers in terms of reducing our potential exposure to fraud. Unlike fruit, vegetables and meat, it is nearly impossible to buy locally or even to reduce the number of steps between farm and fork. The best we can hope to do is buy our spices whole wherever possible and spend £10 to £20 (US$15 to $30) on a spice grinder. It is harder to fake a whole spice and they have a longer shelf life in their whole form.

In addition to buying whole spices, we can switch our seasonings to those that can be more locally sourced. In the UK, this would mean fresh herbs over ground spices. This is a big ask and it would mean the end of curry night, but it might lead to new culinary adventures too. There are so many considerations when it comes to the food we eat – health factors, ethical implications, environmental footprint, cost ... and, somewhere in there, taste! Spices are one more thing (if there's room) to add to our consciousness about food. Just as many people have decided to reduce the amount of meat they eat for environmental or health reasons, perhaps we could consider reducing our dependence on imported spices – even for one meal a week. Who knows, there may even be some beneficial side effects, such as lowering our salt intake, a change our doctors would applaud.

CHAPTER EIGHT

Bogus Beverages

Food fraud is not limited to our solid consumptions, of course – beverages are just as prone. In this chapter we will look at fraud in wine, juice and alcohol. As with other foodstuffs, we see fraud cases with potentially disastrous health implications – the constituents of antifreeze added to sweeten wine, for example. But there are others that hurt only the wallet. Of this latter kind, there is no other area of the food and beverage industry more prone than the wine industry. It is ripe with economic adulteration and this is where we must begin this story.

While some of the stories of food fraud thus far – dung powder in coriander and milk made of urea – may have induced feelings of nausea, stories of fraud from the wine industry may turn your stomach for an entirely different reason. The affluence associated with the high-end wine market is baffling, with rare collections worth millions of pounds, single bottles worth tens of thousands. The amount of money some people are able and willing to spend on some fermented grapes is simply astounding. And of course, where there is such money to be spent, there is also an incentive to rip people off – even if it is done with a sophisticated flair. Some of the stories of adulteration from the fine wine market are worthy of *CSI* or *Sherlock* in their complexity. People who will pay £11,000 (US$16,800) for a good vintage of Romanée-Conti are likely to have a reasonably robust knowledge of the product, or at least to be able to afford the services of someone who does. The criminals who try to swindle these connoisseurs must be equally knowledgeable about wine if they wish to avoid being caught and they must also move in the right circles in order to move their product.

There are resources on both sides, which from our perspective makes for a captivating story.

The best example of these, which is perhaps better known in the US, despite it spanning both sides of the Atlantic, is that of the Jefferson bottles. We shall not go into enormous detail as the story has been told before in articles in the *New Yorker*, the *Independent* and elsewhere. There's even an entire book dedicated to the controversy entitled *The Billionaire's Vinegar*, which was later at the centre of a libel suit. We offer an abridged version here.

A prominent German wine collector, Hardy Rodenstock (legal name Meinhard Görke), claims that in the spring of 1985 he was told about a dozen or so very old bottles of wine that had been uncovered when a wall was taken down in a house in Paris. The bottles were engraved with the initials Th. J. and were all eighteenth-century vintage. Rodenstock acquired the bottles.

Later that year, Rodenstock approached the famous auction house Christie's about selling one of these bottles. The bottle was made of handblown dark green glass and it had no label, but it had 1787 Lafitte (now spelled Lafite) and the Th. J. initials etched into the glass. It was sealed with a thick black wax. Rodenstock claimed that the circumstantial evidence suggested this was a bottle from the personal collection of Thomas Jefferson, third President of the United States.

The experts at Christie's set about authenticating the claim. Glass experts confirmed that the glass and engravings were consistent with late eighteenth-century French style. Historical documents placed Thomas Jefferson in Paris between 1784 and 1789. He had even stayed in Bordeaux in May 1787, visiting many of the wine-producing Châteaux, including Château Lafitte.[1] A letter from 1790, after Jefferson returned to the US, asked that the shipments of Bordeaux wine that he continued to order for himself and President George Washington be marked with his initials. The circumstantial evidence was positive. Michael

Broadbent, who was head of Christie's wine department at the time, sampled two other bottles from the same collection and found them to be authentic based on his extensive knowledge of wine. Christie's sold the bottle in December 1985 to Christopher Forbes, son of Malcolm Forbes and Vice-Chairman of *Forbes* Magazine, for £105,000 (roughly US$156,000 at the time). It was the most expensive bottle of wine ever sold.

As well as the bottle purchased by Forbes and the two sampled by Broadbent, another bottle from the Jefferson collection was purchased by a Middle Eastern businessman. Marvin Shanken, publisher of *Wine Spectator,* purchased a half-bottle of 1784 Château Margaux for a bargain £19,600 (US$30,000). A New York wine merchant had a bottle of 1787 Château Margaux from the collection, which he took to dinner to show off to his friends. A waiter inadvertently knocked the bottle over and the insurance subsequently paid out US$225,000 (£147,300). A German collector named Hans-Peter Frericks also had a bottle of Lafitte he had purchased directly from Rodenstock. US businessman Bill Koch spent half a million dollars on four bottles from the collection: a 1787 Brane-Mouton, a 1784 Brane-Mouton, a 1784 Lafitte and a 1787 Lafitte. That's eleven bottles and seven of them had a combined value of nearly £600,000 (nearly US$1 million).

Frericks became suspicious of his bottle of Lafitte when Sotheby's declined to sell it owing to its uncertain provenance. He sent it to a Munich lab, which examined the contents using radiocarbon dating methods. They looked at the relative proportions of the radioactive isotope ^{14}C in the wine. Radioactive carbon in the atmosphere (in the form of radioactive CO_2) is taken up by plants as they photosynthesise. In the case of wine, when the grapes are picked, they stop acquiring new carbon from the atmosphere and the ^{14}C they have incorporated into their tissues starts to decay. The known rate of decay can then be used to date the sample. Atmospheric nuclear testing, which started in 1945 and continued until the signing of the Limited Test Ban Treaty in 1963, created several tonnes of ^{14}C creating a spike in levels around 1965. The Munich lab concluded that the results of the radiocarbon dating suggested that Frericks' bottle of Lafitte was consistent with organic material from the 1960s or later (meaning the wine had higher levels of ^{14}C than one would expect if it was 200 years old). Frericks sued Rodenstock for selling adulterated wine and won. Rodenstock had another bottle from the collection radiocarbon dated by a Swiss scientist, which showed no similarly high ^{14}C levels. Rodenstock sued Frericks for defamation. They settled out of court.

Bill Koch also began to suspect the authenticity of his half-million-dollar investment in 2005. In preparing to include the bottles as part of an exhibition, Koch's staff couldn't verify their provenance. A curator with the Thomas Jefferson Foundation at Monticello could find no evidence that the bottles ever belonged to Jefferson – no letters or orders within Jefferson's meticulously kept paperwork. Koch, with immense resources at his disposal (*Forbes* estimates his net worth at US$4 billion), began investigating and hired a retired FBI agent, Jim Elroy.

Their investigation revealed titbits of information from Rodenstock's past that brought his character into question. On top of this, Rodenstock had suspiciously good fortune in acquiring incredibly rare wines wherever he travelled.

Elroy took Koch's bottles to French physicist Philippe Hubert, who was using low-level gamma rays to help date wine without having to open the bottle. Much like the radiocarbon dating, the method took advantage of changes in the post-nuclear atmosphere. Prior to the explosion of the first atomic bomb in 1945 there was no radiocaesium (^{137}Cs) in the atmosphere. It's a product of nuclear fission. As with the carbon, the grape vines would take up this element and incorporate it into their tissues, including the grapes. Any wine produced since 1945 will have ^{137}Cs readings while anything produced before then won't. Koch's bottles came back negative. They were at least as old as 1945.

Elroy then brought in the expertise of a retired FBI tool expert and an expert glass engraver to take a closer look at the engravings on the bottle. A journalist had already brought up the fact that Thomas Jefferson tended to use a colon to separate his initials rather than a full point – Th: J. Perhaps the engraving could be the smoking gun. The two experts concluded that the engravings were too uniform to have been made with the technology of the time, a copper wheel operated by a foot pedal. There should be variations in the thickness of the lines, but there wasn't. It was enough for Koch to pursue a civil complaint against Rodenstock. Rodenstock was in Germany and claimed that the US court had no jurisdiction over him there and therefore refused to participate in the case. Koch won a default judgement against Rodenstock in May 2010 for over US$600,000 (£392,000) in damages. Koch later filed a lawsuit against Christie's, but this was thrown out of court as the judge stated that Koch had waited too long to take legal action after he suspected the wine was counterfeit.

Koch continues his crusade against wine swindlers. He has brought experts in to look at his extensive wine cellar and estimates that he has between 400 and 500 fakes among his 12,000-plus bottle collection. In 2013, Koch was awarded damages in a lawsuit he filed in 2007 against one-time billionaire Eric Greenberg. He accused Greenberg of knowingly selling him counterfeit wine – 24 bottles of rare vintage Bordeaux worth US$300,000 (£195,000) – through Zachys auction house in 2005. He was initially awarded US$12.4 million (£8 million) in damages; however, a federal judge later reduced this to just under US$1 million (£650,000). The judge took into account that Koch had already been compensated in a separate case against Zachys and therefore reduced the compensatory damages. And then the judge also reduced the punitive damages from US$12 million (£7.6 million) to US$711,622

(£454,000), stating (and we paraphrase) that this was an overcompensation considering that these actions really only hurt the wallets of billionaires.

Koch also pursued a lawsuit and testified against Rudy Kurniawan, an Indonesian national who was a high-roller in California and a well-known wine aficionado. Kurniawan sold rare vintage wines, except they were actually blends concocted in his kitchen. He used old bottles and mixed old vintages with new to create very believable full-bodied mixtures that he corked, sealed and labelled as authentic. He did this for eight years before his arrest in 2012. The start of his downfall was in 2007 when he tried to sell three magnums of 1982 Château Le Pin through Christie's auction house, but the Château contacted Christie's to say they were fakes. That same year it was discovered that the eight magnums of 1947 Château Lafleur that Kurniawan had sold at auction in 2006 must have been fake as only five magnums of Lafleur were made that year. Then, in 2008, he tried to sell several bottles labelled as Clos Saint-Denis Grand Cru, made by Domaine Ponsot of various vintages between 1945 and 1971. The head of Domaine Ponsot, contacted the auction house to say the Domaine had not made any Clos Saint-Denis prior to 1982. Koch had found that a number of his fake bottles originated from Kurniawan and filed a lawsuit against him in 2009. After an investigation by the FBI Kurniawan was arrested. Even after his trial it remains unclear just how many bottles he sold, turning $100 bottles into $1,000 bottles in his kitchen. Many of his rich and famous customers didn't come forward for the trial, perhaps because they were embarrassed. In 2014, Kurniawan was sentenced to 10 years in prison and ordered to pay over US$48 million (£31.4 million), of which US$28.4 million (£18.3 million) was to compensate his fraud victims (Koch will get $3 million). Kurniawan, once famous for his love of expensive wines, extravagant dinners and jet-setting ways, is now known as the first person ever jailed for selling fake wine in the US.

Koch filed a lawsuit worth millions against New York wine retailer Acker Merrall & Condit for selling him over 200 bottles of counterfeit wine over two years, for which he paid over US$2 million (£1.3 million). The settlement terms were never disclosed other than to say the figure was 'substantial'.

There is a lot of money in the fine and rare wine market. It is low volume (comparably), but there are premium prices. Greenberg, unlike Kurniawan and Rodenstock, was not mixing his own wines; he still made about US$40 million (£26 million) selling wine that was not authentic, though he insists that he wasn't aware of its dubious nature. Estimates of Kurniawan's income from his kitchen mixings range between US$20 million and US$75 million (£11.9 million and £49 million), though he is reportedly poor now. It's unknown how much Rodenstock made on the business.

The string of scandals in the 2000s damaged the fine wine market. Investors no longer trusted what they were buying. Billionaires were feeling burned. Koch was filing lawsuits rather than adding to his wine cellar; he estimates that he has paid out about US$25 million (£16 million) in legal fees. There is money to be made and money to be lost. As of 2015, the market is beginning to recover again.

In all the articles on Rodenstock and Kurniawan, there are stories of grandiose parties and extreme generosity on the part of these hosts. Those who knew them marvelled at their knowledge of wine and sophisticated palates. These men are master mixers, capable of creating a superior product. The contents of the bottle might not have been what the label said, but nobody could deny it was an outstanding wine. Put into the broader context of food fraud where babies are dying in China from melamine in milk, and toxic chemicals are being intentionally added to spices, this really is a case of deceiving the privileged. If only such resources were available to investigate and stop more detrimental forms of food fraud.

Corking copies and salvaging failures

Before we discuss the ways wine has been adulterated through the ages and how we might go about detecting those adulterations, let's just go back to the basics of how wine is made. It's part of understanding the product and therefore asking the right questions when it comes to seeking out the fakes.

There are over 10,000 different grape varieties, which are mostly cultivars of the species *Vitis vinifera* plus a few from *Vitis labrusca*. A varietal is a wine that has been made predominantly from a single grape variety – Cabernet Sauvignon or Pinot Gris, for example. To further complicate matters, there are also clones of many of these grape varieties. The vines accumulate mutations over time, some of which may lead to production of a superior grape or some other desirable trait. This trait may be propagated by taking a cutting of this vine – a clone of the parent – which carries the same DNA and therefore the same mutation. Chardonnay, for example, has at least 34 clonal varieties. For clarification, if a hardwood cutting is taken from a grape vine and grown, it is a clone or clonal variety. If a seed is taken from that same plant, however, and grown, the new vine would be a new variety as it would be genetically different from that plant having had genetic input from a second parent.

So, we start with a grape variety. The fruits of this variety will have slight variations in taste depending on the soil and the unique environmental conditions experienced each season. Once the juice is squeezed from the grapes, it's fermented with the skins or without. The wine is then aged in oak barrels or stainless-steel tanks. These barrels will be made from different oak species depending on the country. The barrels are also toasted – a contained fire is quite literally dropped into the centre of the barrel – which will impart different flavours into the oak. These flavours, such as smoke, spice, vanilla, tobacco and cigar box, will then be passed on to the wine. The wine-makers will request different levels of toasting (light, medium or heavy), depending on the wine being produced. The first fermentation of wine is to convert the sugars to alcohol. Many wines (most reds and some whites) will have a second fermentation, which converts aggressive tart-tasting malic acid into softer creamier-tasting lactic acid. Some wines even undergo a third fermentation. Sparkling wines, for example, undergo a third (and sometimes fourth) fermentation between sugar and yeast to give them their sparkling quality.

In the end, we have a product that is highly variable – from the grapes it is squeezed from, to all the aromatic and flavour compounds that infuse themselves into this liquid along the way. Depending on the origins of this liquid and how it is treated, it may sell for less than £5 (US$7.75) per bottle, up to more than £10,000 (US$15,500) per bottle. This enormous price differential provides an economic incentive for fraud, and the variation in the product can motivate problem-solvers to bend the rules, as you might say. Wine-making is a skilled art form and while some processes always need to happen, it is the artistic tweaks that will lead either to masterpiece or to failure (and a whole lot of high-volume commercial art in between). When nature and human influences work together to create wine and one or the other falls short of the task, there may be a temptation to try and salvage it. And it is this that has motivated many adulterations.

We have learned quite a bit in the 9,000 or so years we have been fermenting grapes, and we know that adulteration to repair less-than-adequate wine was common practice among ancient wine-makers. In Africa, wine would be softened with gypsum or lime. Greeks would add potter's earth, marble dust, salt or sea-water to liven up wine. In the Mediterranean, where it was difficult to keep wines (though it is unclear whether this was due to the climate or simply to a propensity for overindulgence), resin was used to coat the inside of its earthenware containers and was also added to the wine itself to help preserve it. This method is still practised in the form of retsina, the Greek resinated wine. However, these are all arguably just useful additives – creative licence, at worst. There wasn't necessarily any malicious intent or economic objective behind any of them.

Romans added lead to preserve and sweeten sour wine. Lead is sweet (please don't go around finding samples to lick), but it is also a poison. But the Romans didn't know this. It would have simply accumulated in their tissues over time. First it caused abdominal pain and constipation, then their joints and muscles would have begun to ache. They would have a hard time concentrating and would suffer headaches and memory loss. They would lose feeling in their extremities, be unable to speak, lose the ability to procreate and, eventually, die. We didn't really figure out the dangers of lead until the end of the seventeenth century, when lead was still being added to wine. Even after this, lead continued to be used. In eighteenth-century France, tax inspectors became suspicious when they saw the volumes of spoiled wine being brought into Paris. It was legally used to make vinegar, but the volumes coming in did not match the volumes of vinegar going out. Wine merchants were registering as vinegar merchants to bring in the spoiled wine and they were adding litharge (a mineral form of lead oxide) and selling it as proper wine. Addition of a known poison for profit – malicious intent *and* economic fraud.

The devastation of vines in France during the late nineteenth century by the aphid *Phylloxera* was an example of nature falling short in the wine-making process. With almost 2.5 million hectares of vineyards in France destroyed, wiped-out dishonest wine-makers got creative. No doubt even the honest wine-makers were forced onto the path of dishonesty in desperation. Greek raisins were brought in to make raisin wine, though it was not labelled as such. Second, third and fourth pressings of grapes would be made into watery wine and the dye fuchsine, which contains arsenic, would be added to improve its colour. Whether nature was at fault or it was human error, the end result was substandard wine that needed 'fixing'. Stating this on the label would no doubt have turned consumers off. And so adulteration continued, despite numerous laws through the centuries to try and prevent it.

A series of events in the early twentieth century helped restore respect and trust in the wine-making industry. Contributions from scientists such as Louis Pasteur helped to make the art of wine-making more reliable, so there was less need to add ingredients to counter the frequent mishaps. The Appellation Contrôlée system, introduced officially in 1935 in France, helped restrict geographical fraud – only wines from Bordeaux could be labelled as such – but the system also set strict rules for other aspects of wine-making. Furthermore, consumers' palates became more sophisticated and they were less prone to being swindled.

Yet, as we have seen, scandals still happen, and they happen across all wine markets. We have already shared stories of fraud from the fine and rare wine market, concerning wines largely sold at auction houses and through dealers. Prices within this market can be very volatile. At a UK government chemists' conference in 2014, Geoff Taylor, a wine expert with Campden BRI, which provides scientific and technical services to the food and beverage industry, explained this volatility with some hypothetical prices. A 1982 Château Lafite – one of the top five wines in the world – may sell for £19,000 a case (US$29,500); Château Lafite bottled the following year may sell for £3,000 a case (US$4,600). What incentive to change a 3 into a 2! A search on the online database Wine-Searcher for one of the most famous wines in the world, 1990 Romanée-Conti Grand Cru, shows an average price of just over £13,000 per bottle (US$20,000). A bottle of Échezeaux Grand Cru, which is a product of the same wine estate (domaine), is listed at an average price of just under

£900 per bottle (US$1,400). From an analytical viewpoint, these two wines are very similar, yet they fetch very different prices.

There can also be fraud in the mass volume market. These are the bottles that those of us without billions tend to buy that lie in the £5 to £20 (US$7 to US$31) price arena. These are primarily varietal wines. The price differential isn't as great compared with fine wines, but it makes up for it with volume. An Australian red blend, for example, sells for £3.50 (US$5.40) a bottle in the supermarket. Label that as an Australian Shiraz and it is worth double that. There is still money to be made.

At the low-price end of the mass market there are deals we've all seen that seem too good to be true, often in off-licences and corner stores. Three bottles for £10 (US$15)? In 2011 and 2012, wine labelled as Jacob's Creek was being sold at independent off-licences around the UK for as little as £2 (US$3.12). A close look at the back label on the bottle, however, revealed a fatal flaw. It stated that the product was 'Wine of Austrlia'. Ah, the power of typos! Some of the cheap products at these ridiculous prices might not even be wine. They have a low alcohol content and therefore aren't subjected to the same duties. Geoff Taylor from Campden BRI refers to these as aromatised wine products.

Beyond label swapping, there are modern-day scandals with potentially more deleterious effects. In 1985 some Austrian wineries were adding diethylene glycol (a component of antifreeze) to sweeten wine so that it would seem to be the product of late harvest grapes. As with many wines, this wine was exported in bulk and bottled in other countries – largely Germany in this case. To layer one fraud upon another, the German bottlers then illegally mixed the Austrian wine in with German wine so that diethylene glycol ended up in bottles labelled as German wine as well as Austrian wine. Routine standards testing of the German wine revealed the adulteration. Charges and fines were laid in both Germany and Austria and the Austrian wine industry took a severe hit, with many countries banning imports and sales of Austrian wine. The levels of diethylene glycol found in the wine were low, so dozens of bottles of wine would have had to have been consumed to have triggered any symptoms.

The following year (1986), Italy found itself embroiled in a wine scandal. In mid-March people in northeastern Italy began to fall ill. By early April, 17 people had died and 60 had been hospitalised. The cause was wine laced with methyl alcohol, which is also known as methanol and is used largely in the synthesis of other chemicals, including antifreeze, fuel and solvents. It's found naturally in wine in low concentrations, but four large Italian wine producers had added it to boost the alcohol content of some inadequate wines. Testing revealed concentrations 10 times the legally permitted limits. This wine had then been sold on to be bottled for supermarkets. The fraud was revealed when three men who had been drinking large quantities of the same brand of wine died. By the end of the scandal, 24 people had died and Italy's wine export market had plunged. The only silver lining in such incidents is that they prompt countries to tighten their rules and regulations, which Italy did, as did Germany and Austria.

A holistic approach to wine crime

So how do we find the fakes? The short answer is that it is complicated. It requires a very holistic approach, where product knowledge is essential – far more so than with any other food or beverage.

In terms of scientific analyses, there are sophisticated tests that can be used to look at the volatile compounds, aroma components, minerals and trace elements in the wine. However, many of these can be challenged in a fraud case because so much can be introduced to a wine in its normal (and legitimate) evolution, while other markers can disappear with time as the wine continues to evolve.

Analysis of the carbon, hydrogen and oxygen stable isotope ratios from the water and ethanol in the wine can be used to characterise the year of production of a legally defined wine region (appellation) as they are heavily influenced by climate. Combine this stable isotopic analysis with a measure of the trace elements in the wine and it becomes possible to trace the wine back to the soil from a particular slope in a sub-region of Burgundy. This type of analysis is critical to ensuring that the strict laws and regulations governing appellations are being followed.

As yet, there is no varietal test for wine, no dipstick that can be plunged into the bottle that confirms it is made with Cabernet Sauvignon grapes. Many attempts have been made. Scientists have tried to salvage residual DNA and use markers to identify the different cultivars, with little success. There has been considerable work in Australia to look at the feasibility of using visible near-infrared spectroscopy (discussed in the last chapter) for authenticity testing. They are trying to establish signatures that distinguish Spanish Tempranillo from Australian Tempranillo,[2] for example, or that discriminate between Australian Chardonnay and Australian Riesling.[3] However, there is limited wider application of these tests at this point as the studies have focused on specific wines from specific areas.

Sensory testing – as we described in the section on olive oil in Chapter 3 – remains one of the more reliable methods of testing wine. Such testing is generally used in quality control when identifying defects in the wine, such as mustiness that is attributed to the presence of trichloro-anisole (TCA). Sensory testing would be unlikely to stand up in court in an authenticity trial, however.

The big problem with all of these tests? You need to open the wine. As you can imagine, this is not ideal for someone trying to authenticate a rare vintage they've just paid a small fortune for. It is for this reason, probably more than any other, that most suspicions of fraud have been raised on everything BUT the actual contents of the bottle.

Sometimes, it's a matter of numbers that simply don't add up. The amount of soured wine into Paris versus the amount of vinegar out. The number of magnums of 1947 Château Lafleur produced versus the number sold. In 2000, an Italian distributor of Sassicaia – a prized Italian wine worth over £130 (US$201) a bottle – became suspicious when he noticed a plethora of 1994 and 1995 vintages out on the market. The Guardia di Finanza, Italy's fraud squad, which deals with all types of fraud, stepped in and investigated. They traced the fakes back to Neapolitan counterfeiters in Tuscany with a known history of fraud. The labels on the real bottles were embossed and had more intense colours than the fakes. The Italian police confiscated 16,000 bottles of fake Sassicaia and arrested two people in connection with the counterfeiting scam.

Taylor (from Campden BRI) has been called in on a number of suspected fraud cases. He uses chemical analyses in his work, but he also advocates the need for product knowledge and watchful eyes. He shared a case study with the audience at the government chemists' conference:

This was for a famous champagne house as they were concerned that some of their products were being copied. Fortunately, this house produced wines in a certain style. Champagne undergoes a primary fermentation that creates a base wine, this base wine is then bottled with sugar and yeast and undergoes a secondary fermentation in bottle. Some champagne houses modify their base wine by putting it through malolactic fermentation [to make it taste softer and creamier]. However, this particular champagne house does not put any of their product through malolactic fermentation. This is where product knowledge is essential. This knowledge allowed us to specifically look for lactic acid as an indicator that this was not their product. This was one bullet. The second [bullet] was that champagne is bottled with very good corks and closures. Typically, it would be a natural cork with two to three rondels [cork disks] in contact with the surface

*of the liquid. These rondels are some of the purest grade cork
you can get. This champagne house always seals with two
to three rondels depending on the quality of the champagne.
The closure in the suspicious bottle used an agglomerate cork
[the cork equivalent of chip or particle board] and only one
rondel. We used a combination of analytical chemistry and
observation.*

In another case, Taylor used the presence of the reverse
epsilon (ɘ) among the markings in the glass of the wine
bottle to show the bottle was a fake. The reverse epsilon
only started being used on bottles in the 1980s, yet the
bottle was supposedly of an early twentieth-century
vintage. The glass itself was also flawless, which is
uncharacteristic of glass from this period.

Some wines are not allowed to be bulk exported and
this can be a clue that there's been some deception. Rioja,
for example, has to be bottled in Spain, so you should never
find a bottle that has any markers that indicate it has been
imported and bottled by a coded company starting with
W as that would indicate a UK bottler. Finding wine fraud
truly is a matter of holistic super-sleuthing.

On the whole, however, the wine industry is very highly
regulated with laws specifically designed to combat fraud.
Vintners take great pride in their products and they want to
protect their reputation. As consumers, we can only arm
ourselves with knowledge. The first principle is that you
generally get what you pay for. Be suspicious of bottles
under £5 (US\$8). If they are indeed wine, they may not
cause you great harm, but it is likely to be an assault on
your palate. If you are lucky enough to be in the fine and
rare wine market, then it is an investment and as such it is
worthy of some research. If you don't have the knowledge,
spend a little more to bring in someone who does. For
everyone else, read the label, look for typos and other
suspicious details and, most importantly, when you find
something you like ... buy a case!

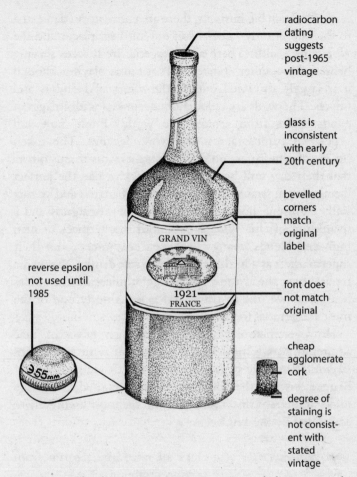

radiocarbon dating suggests post-1965 vintage

glass is inconsistent with early 20th century

bevelled corners match original label

GRAND VIN

1921
FRANCE

reverse epsilon not used until 1985

font does not match original

cheap agglomerate cork

degree of staining is not consistent with stated vintage

Figure 8.1. Sleuthing suspicious wine requires a holistic approach, often in lieu of chemical analyses.

Getting squeezed

The juice market in the US is worth an estimated US$33 billion (£21 billion) per year. Every few years there seems to be another juice craze hitting the media, boasting the antioxidant powers of this juice or the vitamin boost of that juice. It is such big business, there are entire stores dedicated to the sale of only juices. Visit one of these places and the vibe is of health. There are energetic, fresh faces serving behind the counter. There's upbeat music playing, though it is largely drowned out by the whirr of blenders and juicers. The walls are colourfully decorated with images of exotic fruits from around the world. Fresh fruit and vegetables tumble across the work benches. The token square of wheatgrass on the counter screams to customers that their day will be better once they've had the perfect blend of açai, strawberries, raspberries, bananas and yogurt with a ginseng boost, a double shot of wheatgrass and a sprinkling of chia. Unable to afford the luxuries of such high-end blends, many consumers buy juices from their supermarkets as a healthier choice to soft drinks. Yet media reports over the last few years suggest juice may not be as healthy as we like to think and it is certainly one of the most adulterated foodstuffs on the market.

However, before we get into the nitty-gritty of juice adulteration, we must first acquaint you with some juice jargon. When we began researching this section, we got bogged down in the different terminology used to describe juice. So, here is the crash course on the most useful terms, as defined by the FAO:

100% fruit juice	— it's all juice and it's not from concentrate.
Fresh squeezed	— it's not pasteurised and will be found in the refrigerated section of the supermarket.
Juice concentrate	— the water has been removed from the juice. This makes it more economical to transport and store.
Not from concentrate	— single strength juice (it hasn't been concentrated) and it is usually pasteurised.

From concentrate   —   juice that's been reconstituted
                            from concentrate and then
                            pasteurised.

Juice blend   —   a mixture of pure juices.

Puree   —   contains pulp and is more
                            viscous than juice.

Nectar   —   it is 25 to 50 per cent juice with
                            added water, sugar and acid.

Juice drinks   —   it contains 10 to 20 per cent
                            juice (but standards differ by
                            country).

Let's consider the most popular of the juices: the modest
orange. The US alone squeezes around 30 billion oranges
annually for their vitamin C-rich juices and it accounts for
over half of all juice consumption. Its popularity alone
makes it a target for fraud. The most common adulterants
of orange juice are water, sugar, pulp wash (which is, as the
name suggest, a water wash of the previously pressed pulp
to try and squeeze out any remaining juice), and less
expensive juices. These are, for the most part, rather
innocuous adulterants compared with some we've shared
within the pages of this book.

The addition of water, particularly when reconstituting
juices from concentrate, is a straightforward and easy
method of making juice stretch a little further. However, it
is also an easy adulterant to detect. The brix measurement,
taken with a refractometer, gives an approximation of the
total sugar content; $1°$brix is equivalent to 1g of sucrose in
100g of solution. Simply adding water can reduce the brix
measurement of the juice and the Codex Alimentarius
gives strict guidance on acceptable brix values for every
type of juice on the market. This forces the fraudsters to
add a second adulterant, sugar.

This is where things become somewhat less innocuous.
First of all, the fruit we consume has been bred into
unnaturally sweet proportions. We have selected species

bearing the sweetest fruits because they taste best. Then, over centuries, we have cultivated the plants within that species that have the sweetest product. The result is fruit with a higher sugar content than what might be harvested from the wild. The higher sugar content in fruit has been noticed by zookeepers who have fed their animals, mostly primates, human-grade fruit on a daily basis. The animals were experiencing anxious and unsettled behaviour – pacing, inability to concentrate, aggression. Does this sound familiar to any primary school teachers? When they cut the quantity of fruit back and served them vegetables instead, the animals settled right down. They have attributed this to the higher sugar content in the fruits we humans like to eat. Juices take these already sweet fruits and concentrate the sugars by removing the pulp and fibre. Add to this the fraudster who is trying to mask some added water by adding more sugar to the juice, and suddenly this doesn't seem like a very healthy product. A 250ml (8.4fl oz) serving of pomegranate, blueberry and açaí smoothie can contain the equivalent amount of sugar of a 330ml (11fl oz) can of Coke.

Any sugar can be added to hide water adulteration and many have been tried, including cane, corn and beet sugar. HFCS, which we introduced in Chapter 2 with respect to honey, is another cheap sweetener that's being added to juice (as well as many other foods). HFCS is cheaper than cane sugar, contains the same amount of calories, yet is processed differently by the body. Research to date has indicated that even after just two weeks of eating food with HFCS daily, risk factors associated with heart disease are increased and there may also be evidence of increased risk of diabetes and liver damage. Luckily, both sugar cane and corn are C_4 plants and so stable isotope analysis, as we've described previously, can be used to distinguish whether these sugars have been added to the juice. This has made beet sugar (a C_3 plant) the first-choice adulterant when it comes to juice.

To detect beet sugar in orange juice, stable isotope signatures are still useful, but instead of looking at carbon, analysts look at the isotope ratios of hydrogen and oxygen. The heavier isotopes of hydrogen and oxygen are found in higher proportions in the groundwater of warm citrus-producing regions compared with the groundwater of colder regions where sugar beets are grown; it's back to the concept of isoscapes. Researchers from the Universities of Miami and California examined sucrose isolated from oranges and compared it with sucrose isolated from beets. They could easily distinguish the two sources of sucrose based on their isotope ratios.[4] In fact, because oranges are reasonably restricted as to where they can grow, the isotope signatures were very consistent between all of the samples of orange juice examined. The researchers could predict with 99.99 per cent confidence that the isotope ratios for sucrose extracted from orange juice would fall within a particular range. The isotope ratios for beets are a little more variable, and this is because the geographical range where they can be grown is somewhat greater – beets grown in colder regions are depleted in heavier isotopes, whereas beets grown in warmer regions are less so. This means sugar from beets grown in colder regions will be more easily detected in orange juice than sugar from warm-weather beets. So in a strange turn of events, the fraudsters might need to verify the origins of their beet sugar supply chain in order to reduce their chances of being caught. Hopefully we have not just armed the fraudsters with information they didn't already have!

As well as being adulterated with water and sugar, cheaper juices can be added to make more expensive juices stretch further; in the case of orange juice, this is often grapefruit juice. As luck would have it, these two citrus fruits have different flavonoids (plant pigments). As an aside, the major flavonoid in grapefruit is naringin, which not only gives the fruit its bitter taste, but is also the compound that can interfere with some prescription drugs, usually by increasing or decreasing how the drugs are absorbed in the intestine. Hesperidin is the major flavonoid in oranges. HPLC can be used to separate the compounds and therefore determine whether grapefruit juice has been added to orange. However, another method, which is not new, has recently been applied to juice authentication: front-face fluorescence spectroscopy. In a *very* simplified explanation, light from a xenon lamp is beamed at a sample of the cloudy fruit juice at incremental wavelengths through a filter. Some of the light is absorbed by the sample (as in other methods of spectroscopy) and this gets the molecules all excited in the juice. When certain molecules get excited by the light radiation they fluoresce (emit light), usually at a lower energy (longer wavelength) than what they've absorbed. This light gets scattered in all directions and is measured by a detector that is placed where it will get the least interference from the excitation light beam. The spectra for orange juice and grapefruit juice have different fluorophores (fluorescing compounds), which are probably the flavonoids again. HPLC and other spectroscopy techniques are used to screen various types of juice for the presence of cheaper juices and they all work on the same principle of identifying distinct chemical compounds that are unique to particular groups of plants.

The addition of pulp wash is another common practice. After the first squeeze of juice from the pulp, juice vesicles remain that still contain juice. The separated pulp is washed in a way that recovers about 90 per cent of the juice from these vesicles. This juice is paler and more bitter and for this reason is considered a lower quality product – it isn't considered to be juice. In the US, the pulp wash can be put back into frozen concentrated juice that is being made from the same oranges, but not any other type of juice. If it's not put back into the juice as they are making it, the pulp wash can be concentrated and packaged up to be sold as fruit solids or pulp wash, which is a less expensive base for fruit drinks (not to be confused with fruit juices).

Manufacturers are not allowed to add pulp wash that isn't from the oranges they are squeezing right there and then into any other type of orange juice. But it happens, usually along with the addition of citric acid, sugar, amino acids and even trace metals to try and replicate the chemical profile of pure orange juice. So how do you tell the first-squeeze juice from the really super-squeezed wash? Research has been done to try and find that magic chemical compound that can easily distinguish authentic juice from juice that has had pulp wash added. Dimethylproline (an analogue of the amino acid proline) is potentially a good candidate as it appears more prominently in the chemical signature of pulp wash.[5] However, the US FDA has gone one step further, and in Florida all pulp wash has a trace compound introduced into the pulp wash. Much as chicken waste is dyed to make sure it doesn't re-enter the human food chain, Florida adds small amounts of sodium benzoate (an approved food preservative – E211 in Europe) to pulp wash. The preservative is not allowed in orange juice and can easily be detected using HPLC. If it's present, it indicates that pulp wash has been added to the juice. Though one may question whether the preservative added to detect the fraud is potentially more harmful than having a little pulp wash in your juice.

There is also considerable mislabelling of fruit juices (and this is where the juice jargon comes in handy). Reconstituted juice (from concentrate) may be labelled as fresh squeezed as it's generally worth more, for example. Gas chromatography has been used to analyse volatile compounds in juices with some success, as the way each of the juices is treated – pasteurisation for example – will affect these compounds and help differentiate the fresh squeezed from the reconstituted. However, we have mentioned before that there is considerable natural variability of foodstuffs and orange juice is no different.

At least 16 different compounds have to be analysed in order to develop a chemical signature for the different types of juices as there is enough variability between all of them that no one compound alone is enough to differentiate them. Even with straightforward tests, such types of fraud, which pose no health risks whatsoever, are not likely ever to be a priority. The best way to avoid this type of fraud is to just eat a piece of fruit, and gain the additional advantage of fibre!

Yet some adulterations of juice are not as risk-free. Many have undeclared additives that could potentially pose a risk to human health. In 2011, sports drinks and juices were in the headlines, as illegal clouding agents were being used by Taiwanese manufacturers of these beverages. Clouding agents are legitimate food additives that help maintain a uniform emulsion of the juice – in other words, they make them appear cloudy. Palm oil and citrus are both natural clouding agents that are commonly used in the food industry. However, in 2011, it was discovered that clouding agent manufacturers in Taiwan had been using the chemical di(2-ethylhexyl) phthalate (DEHP) in their products, not only because it was cheaper, but also because it had a preserving effect compared with other clouding agents. DEHP is a plasticiser, which is used to make plastics, such as polyvinyl chloride (PVC), more flexible. According to reports, the plasticisers had been making their way into foodstuffs such as sports drinks, fruit juice, tea drinks, fruit jam or jelly and (ironically) health food supplements for years and had gone unnoticed. DEHP, like several other phthalates, are endocrine disrupters – they adversely affect sex and thyroid hormones, reproductive function and neurodevelopment.

The scandal broke when routine testing by Taiwan's Food and Drug Administration (TFDA) discovered DEHP in a probiotic supplement, quite by accident. They started to find it in a diversity of other products that didn't immediately appear to have anything in common. They detected DEHP in 20,880kg (46,000lb) of juice, fruit jam, syrup, fruit powder and yogurt powder, nearly half a million bottles of sports drinks, flavoured juices and artificial tea drinks, and 133,887 boxes of probiotic powders.[6] Taiwan's Department of Health (DOH) started to search for the ingredient common to all these products, and eventually all roads led to a perfumery-chemical company that was a supplier of emulsifiers. By the end of the month, the DOH had identified a second perfumery-chemical company that had been using another phthalate, diisononyl phthalate (DINP), as a clouding agent in its products for over a decade. Together, the two companies had supplied over 119 other companies, which had used the clouding agents in 371 products. The products had been exported to around 22 countries, including Australia, mainland Europe, the UK, Canada, New Zealand, Brazil and Japan. Taiwan had a crisis on its hands. The Taiwan government informed the WHO and other international authorities. Within a month the government required that all food products in these categories have a phthalate-free certificate, which was obtained by submitting product samples for analysis at accredited laboratories. No product could be exported without this certificate, nor could it be put on supermarket shelves. The rule was rescinded once it became clear that the plasticisers were no longer being found in food products.

The Taiwan phthalate scandal was compared with the Chinese melamine scandal, but experts have said that the Taiwan government acted more swiftly and more openly than the Chinese government. The Taiwan government sent communications out daily at the height of the scandal. In the wake of the scandal, they imposed stricter penalties for those found adulterating food with potentially harmful chemicals. They also put resources into clinics around the country that people potentially affected by the products could attend for assessment and additional health information. Finally, the government started epidemiological studies to examine those that had been exposed to the phthalates.

For those of us who have squeezed our own juice, we know that within hours after making it, the juice starts to change, even in the fridge. Oxidation starts to turn the juice brown and the flavour starts to change as fermentation begins. It is therefore concerning that leading brands of orange juice sold in the UK, which list freshly squeezed oranges as the only ingredient, manage to look and taste the same even after they've been in the fridge for a week. What isn't being declared on the label? How can these differ so much from the juice we squeeze at home? Something is amiss.

The booze bamboozle

As with wine and oil, spirits have a long history of adulteration. And if the health impacts of the alcohol itself aren't questionable enough, the substances it has been adulterated with over the years certainly are. Turpentine (a volatile oil used in paints and varnishes), sulphuric acid (a corrosive acid used in processing minerals and manufacturing fertilisers), chloroform (the stuff villains in mystery movies douse a handkerchief with to knock out their victims), isopropyl alcohol and acetone (both solvents) have all been added to spirits for one reason or another with the common motive of making some money.

The most adulterated and counterfeited spirit products appear to be those that are relatively tasteless and clear – vodka and gin, for example. In 2011 and 2012, the UK FSA revealed that they were seeing an increase in the amount of counterfeit alcohol, particularly vodka. In fact, fake alcohol constituted nearly half (45 per cent) of the 170 fraud cases in the FSA food fraud database in 2012. Raids carried out in 2011 in Lincolnshire, England resulted in 88 litres of fake vodka being seized that were made with isopropyl alcohol, which is cheap and causes rapid intoxication. The alcohol was being made in dodgy facilities using unhygienic practices, though this seems like a rather minor point when they are quite literally bottling poison for sale. Symptoms of isopropyl alcohol poisoning include dizziness, low blood pressure, abdominal pain, vomiting and coma, but it can also lead to kidney failure. The symptoms are so acute that the people experiencing them believe that they've had a drug slipped into their drink, when in fact it's the counterfeit alcohol at work.

Unflavoured vodka is composed of ethanol, water and sometimes small amounts of sugar. Counterfeit versions can be quickly identified from their chemical signature as there are always impurities that have been picked up from the tubing or vessels that are being used for its fabrication, which aren't seen in pure vodka. The counterfeit vodka tends to be in the 'too good to be true' price range and with unknown brand labels. Again, you get what you pay for. Even in vodka's native land of Russia, fakes have been found. Though vodka is already cheap in Russia, many people prefer home-made varieties because they cost even less. In 2005, 25 people were killed by a batch of illegal vodka that contained methanol. The cost savings simply aren't worth it.

While we have seen that we are vulnerable to fraud even as alert consumers in the supermarket, imagine the potential for swindling the intoxicated fun-seeker. We can clearly make a decision not to buy the cheap bottle of gin, but when we are in a bar, a nightclub or even a restaurant we are introducing another dimension to the supply chain at a time when we are perhaps a little less alert to such manipulations.

Other alcohols that are not as easily replicated in terms of appearance and flavour may be subject to false labelling, particularly premium spirits. A cheap whisky blend, for example, might be poured into an 18-year-old single malt Talisker bottle. Or more likely, a cheaper brand is served in lieu of the more expensive one ordered at a restaurant or bar. Though this is certainly less of a health risk than the counterfeit vodka, it is still a crime. For this type of fraud, flavour and aroma compounds, which are often unique to each distillery, can be used to determine whether the beverage originated from there. The carbon isotope ratios can also be used to differentiate between whiskies as these vary more between the products of different distilleries than they do between production years at the same distillery.

A counterfeit Scotch operation in India was unearthed in 2015. Used bottles of Glenlivet, Glenfiddich, Ballantine's and other known brands were being bought from scrap dealers and filled with cheap Indian-made whisky. Restaurants and bars are supposed to break the empty bottles of these big brands to prevent this from happening, but some clearly don't. Some individuals will earn a little extra money by selling empty bottles to scrap dealers. As we've mentioned before, such criminal rackets require a certain unspoken cooperation between the parties. Surely the scrap dealers that collect the bottles know what the empties are being used for. Surely the restaurants and bars that buy the bottles from the 'distributors' must also be suspicious of a bottle of Glenfiddich that's selling for half the normal price. Yet nobody says anything because they are all benefiting in some way.

The Scotch Whisky Association (SWA) has stated that millions of litres of this fake stuff is even finding its way back into Europe. Between 2009 and 2014, plans to export approximately 4.5 million litres (just under one million gallons) of fake whisky into Europe were thwarted by the authorities. This is just what was actually found. There aren't any estimates as to how many fakes actually made it into Europe. In 2013, the SWA began lodging trademark infringements in more than 30 countries; 19 objections were lodged in India alone. It is legitimate producers that suffer as a result of counterfeiting and so they are forced to take action to protect their product. Protecting the authenticity of the name Scotch Whisky is one way to do this. As of 2014, any producers, blenders, bottlers, labellers or bulk importers of Scotch Whisky must apply to HM Revenue and Customs to be verified under the Spirit Drinks Verification Scheme. About a quarter of all UK food and drink exports are Scotch Whisky products – so the UK government has to support the industry by cracking down on the fraudsters.

So sadly not even our beverages are safe from the fraudsters. Whether we are buying rare vintage wines from Christie's or just pouring a glass of OJ for breakfast, we are vulnerable to bogus beverages. Just as with other foodstuffs, we can take steps as consumers to reduce our susceptibility to such frauds. When it comes to cheap wine and cheap booze, you get what you pay for. Even if they may not technically be counterfeit mixtures of different solvents, they may not be far from it in terms of their taste. Avoid them and familiarise yourself with the symptoms of alcohol poisoning. Last, but not least, eat fruit.

CHAPTER NINE
White Collard Crimes

We are not yet at risk of finding fake tomatoes or spinach on our supermarket shelves, thank goodness. Though, if the writers are honest, we didn't know about fake eggs or grapes (from the introduction) prior to doing research for this book, so who knows what might be possible with the right ingredients! Yet we still find fraud among our wholesome fruit, vegetables and grains.

In 2015, boxes of mandarins were found in a Chinese wholesale market. In clear, bold green lettering across the side of the box it said 'AUSTRALIA', and then below, in smaller lettering, 'sweet fruit'. The image on the side of the box was of a muted, somewhat out-of-focus grassland. And in the foreground, perfectly in focus, two juvenile ... lions? What lions have to do with oranges or Australia is unclear and it is for this reason the boxes were investigated. They were, of course, a product of China.

This was just one of many examples of Chinese citrus being rebranded as Australian citrus because the Australian products are more expensive. The imported oranges carry a premium largely because they are viewed as 'clean' fruit. Between 2010 and 2012, numerous stories came out that a number of domestic oranges were being sold in China that had had their peel rubbed in carcinogenic Sudan Red dye to make them appear riper than they actually were. The fruit had been picked early in order to get a jump on competitors ahead of the domestic citrus season. It was either artificially matured or dyed to look ripe and sold as US or Australian produce. Because, of course, Asian fruit wasn't in season yet and consumers knew that. While the dye didn't appear to affect the flesh, there is obviously a strong likelihood that the dye can be inadvertently ingested during the process of peeling the fruit. No fancy test was necessary as the dye simply rubbed off in your hand or, preferably, on a cloth or tissue. Once the scandal broke all was revealed and consumers lost confidence in their domestic citrus fruits.

Perhaps as a result of the citrus scandals, Australia's exports to China have been on a steady increase; their citrus exports are currently valued at around £19.3 million (US$30 million). Yet Australian producers are wary, as they obviously don't want to see their packaging being replicated for the purposes of deceiving customers into thinking that they're buying Australian when they are not. This is further complicated by the fact that some of this misleading produce is being sold on to other Asian countries. It is tarnishing the Australian brand.

In 2014, Hong Kong had its first suspicious citrus scandal when 5,200 oranges that had been given fake Sunkist brand labels were found on market stalls in Yuen Long. As well as the oranges, the authorities seized over 100,000 forged Sunkist labels. The fakes were uncovered when a customer noticed that the fruit didn't taste the same and that the peel was thicker than normal. The complaint led to a two-week investigation that culminated in the arrest of the owner of the stalls along with three sales staff.

Fruits and vegetables, in their whole form, generally do not make an appearance in the food fraud databases. Squeeze the oil or juice out of them and it's fair game, but it would be rather bold to try and dye a parsnip orange and sell it as a carrot. Yet as we saw with the citrus example, there are ways we can still be deceived by the fruit and vegetables we buy – whether it's how fresh they are, where they've come from or how they were produced.

The preservation principle

As anyone who has ever had the privilege of growing their own fruits and vegetables knows, there are natural gluts and lulls in the growing period. You can be desperate for anything that's not a root vegetable during the winter months and buried beneath more berries than you know what to do with during the summer. It is, therefore, common sense to find ways to preserve things so that you can extend the season in which the garden's bounty can be enjoyed. Food science has helped us do this.

While many of these methods keep the fruit or vegetable relatively intact – blanching and freezing, or pickling, for example – others turn it into less recognisable products, and this is where the fraudsters have some operating room. Purees, for example, which can be frozen or canned (often for baby food), have been prone to some species swapping and blending. Apricot puree has been substituted for pumpkin puree, apples and plums have been worked into raspberry purees and the purees of cheaper fruits have made their way into more expensive jams. The purees may be an end product themselves, but more often than not they are used in other products such as pudding, yogurt, ice cream and baked goods. In this way the fraud gets passed along the food chain.

To detect these adulterations a number of methods have been used. PCR has been used to differentiate the species in a puree sample based on the DNA present. Chromatography techniques have also been used to identify signature compounds that are unique to particular plant species as a way of identifying what's in the macerated mixture.

Of course, we are not just faced with trying to preserve our fruits and vegetables for leaner times. In this global food network we have created, we must also find ways of transporting these (often) soft and fleshy products thousands of miles with as little spoilage as possible. To do this, many fruits and vegetables are picked before they are ripe. This is fine for certain groups of plants, known as climacteric plants, because they continue to ripen even once they are picked. These plants produce a lot of ethylene and undergo rapid respiration as they ripen; examples include apples, avocados, apricots, bananas, mangoes, peaches and tomatoes. Non-climacteric plants, as you would expect, do not continue to ripen once they are picked and these include bell peppers, cherries, citrus fruits, cucumbers, grapes, strawberries and pineapples. When was the last time you saw a pineapple that wasn't green in a UK supermarket?

This basic difference in plant physiology can be used to the advantage of the food industry. Bananas, for example, which would have very little tolerance for being shipped about when perfectly ripe, are instead picked hard and green and stored mature but unripe. When they are needed, the bananas are ripened in a room with ethylene pumped into it, which speeds up the ripening process. Ethylene can also be used to de-green non-climacteric fruits, such as citrus. It will change the colour of the peel from green to orange or yellow so that the fruit appears ripe, but unlike the effect on bananas, it won't actually help the fruit to ripen. Generally, non-climacteric fruits have to be picked as ripe as possible and then refrigeration or other methods must be used to slow the rotting process.

It is here, in this attempt to slow the decay of our fruit and veg, that we can sometimes be deceived. The most heinous example of this comes from Bangladesh. In 2014, the Bangladesh government started to crack down on the illegal use of formalin in preserving fruit. Formalin is formaldehyde gas dissolved in water and is well known as the substance used to preserve human bodies as well as biological specimens. However, formaldehyde is also found naturally as it is an organic compound produced as a consequence of biological processes. Humans produce it, as do fruiting plants. But the Bangladesh authorities were measuring formaldehyde levels to more than 1,500 times the natural levels found in fruits. Mangoes, for example, were being picked and sprayed with formalin to keep them fresh prior to shipment. When traders received the mangoes, they were getting rid of any that hadn't survived the journey and then spraying the rest of the mangoes again with formalin before sending them off to retailers. Sometimes retailers were giving the mangoes a third spray of formalin before displaying them on their shelves.

Inhaling formaldehyde can cause irritation of the respiratory system and the eyes as well as nausea, dizziness and headaches. Long-term exposure to formaldehyde has been linked with some forms of cancer. The effects of consuming formalin are less well known, though as it occurs naturally we are eating it as part of our daily diet already. WHO has set the tolerable daily intake level for formaldehyde at 0.15mg/kg per day. This means that a 70kg (154lb) individual shouldn't consume more than 10.5mg (0.37oz) of formaldehyde daily. Some of the mangoes in Bangladesh were measured as having 46 parts per million (ppm) formaldehyde, so it would take more than one mango (around 250g (8.8oz) of mango flesh) for that person to reach this limit.

The use of formalin is, of course, illegal and if perpetrators are found guilty of using it, they can receive a maximum sentence of life in prison. Yet it is still used. Luckily, it is one of the easier types of fraud to detect. A hand-held device can be purchased for anywhere from £30 to £300 (US$46 to US$467), which will give a reading, no doubt with varying degrees of accuracy, of the amount of formaldehyde in the air. However, there are also formaldehyde dip detectors that are perhaps more accurate for determining formaldehyde concentrations in liquids and solids. These are little strips of paper that simply need to be held to the product being tested. The indicator paper will change colour and this can be compared with a reference chart to give a value in ppm. It's very similar to a pH test. Thirty strips can be bought online for around £40 (US$65) and this is as straightforward as tests come.

While the use of formalin is clearly an illegal practice, there are methods of preservation that are perfectly legal within the food industry, but perhaps no less deceitful. There are patented formulations that are classified as processing aids, which can help inhibit the oxidation of fruit and vegetables. All of us have used such methods to some extent − we have sprinkled some lemon juice on apple slices to stop them going brown or kept carrot sticks in water for the same purpose. These patented processing aids, which may be no less natural in their origins than some lemon juice, can prevent a cut-up apple from turning brown for approximately 21 days. They do not need to be disclosed on any food label because they are considered a processing aid rather than an ingredient. They don't pose any health risk as formalin does, but something seems deceitful about it, perhaps because most of us (ourselves included prior to writing this book) aren't aware that there are methods to keep an apple from turning brown for that long. As a result, we make an assumption that the sliced apple is reasonably freshly packaged − perhaps within a day or two. And this is where food processing begins to blur the lines between what is real and what isn't.

Over the last decade, the use of nanoparticles in food has been a hot research area, particularly for preservation. We're not talking about food packaging here, we mean *in* our food. Silver nanoparticles, for example, have antimicrobial properties and are used in many products where one might not want microbial growth − our underpants for instance. Silver nanoparticles have therefore been used as pesticides but are also being used in coatings for fruit and vegetables to preserve their freshness. Asparagus spears, for example, can be kept in good quality for up to 25 days in cold storage when coated with silver nanoparticles, compared with only 15 days when not. Studies have shown that even after repeated washing, the silver nanoparticles are still found on the skin of some products and, owing to their size, can penetrate the skin and enter the flesh.[1]

The impact of eating these nanomaterials is unclear and it is a hotly debated topic. Nanoparticles are small enough to pass through cell membranes and the science about how these nanoparticles spread out in a living body and accumulate in different tissues is just starting to unfold. Nanoparticles can also cross the placental barrier, raising concerns about how they may affect the development of an unborn baby. Even as short-term studies are being published, it's too early to understand any long-term effects of the chronic consumption of nanoparticles. There are clearly advantageous applications of nanotechnology in the food industry. Nano-sized salt, for instance, is one-thousandth the size of a normal grain of salt, but with one million times the surface area. This means far less is required to get the same level of seasoning, which is ideal for reducing salt intake. However, the science on the impact of nanomaterials in our food needs to catch up with the applications.

In the US, the use of nanoparticles in food is subject to the same regulations as adding any other ingredient to food. However, at the moment, the US does not require nano-sized ingredients to be labelled. While the EU does require any nano-sized ingredients to be labelled as such on the packaging, it doesn't require any nanomaterials used in processing to be declared on the label. So, does a silver nanoparticle coating that migrates into the flesh of a fruit count as a process or an ingredient? This is clearly a grey area. In the meantime, there is a flurry of research to develop the best methods for detecting and quantifying nanoparticles in food.

The differentiation of processing methods and ingredients is an important one. There are 6,000 food additives, including flavourings, glazing agents and improvers, which are used in the behind-the-scenes processing of our food. Remember the CGI analogy from the introduction? These processing methods can convert an immature cheese into a mature cheese in 72 hours. The consumer is under the illusion that they are getting a wedge from a round of a carefully aged cheese – a cave-aged cheddar perhaps – when in fact it's a stunt cheese disguised by some fancy processing effects. What goes on behind the scenes legally and what is omitted from the label is a topic for many more books. Felicity Lawrence's *Not on the Label* and Joanna Blythman's *Swallow This* both cover these topics extensively.

But does this constitute food fraud? The EU has yet to develop an official definition of food fraud, but a definition from the US includes 'the deliberate and intentional substitution, addition, tampering, or misrepresentation of food, food ingredients, or food packaging; or false or misleading statements made about a product, for economic gain'.[2] Whether a sliced apple that's 21 days old and white as the day it was cut is a misrepresentation of food will require expert interpretation of existing food laws. In our non-expert opinion, it feels as though not all the information is being disclosed and it therefore has an air of dishonesty about it.

On the origin of veggies

As we saw with the Asian citrus fruits being sold as a product of Australia, making claims about where fruits and vegetables originate is one of the ways fraudsters can make a little extra money. Tropea red onion (*Allium cepa* var. Tropea), for example, may look like a normal (albeit slightly elongated) red onion to most consumers. Yet it is a highly prized variety that is cultivated in a very small area in Calabria, a region of southern Italy. Owing to the reputation of the onions grown in this region, this variety was awarded protected geographical indication (PGI) certification under the name 'Cipolla Rossa di Tropea Calabria' by the EU in 2008. While the designation gives consumers assurance about the origin and quality of the onion, it also gives fraudsters an opportunity to cash in on a premium product. In 2008, it was estimated that just over 18,000 tonnes of these red onions were produced in the region, yet just over 90,000 tonnes of onions labelled under this PGI brand were sold. As we have seen in previous examples, the numbers simply don't add up.

To try to protect the PGI designation, researchers began working on some simple tests that could be performed to authenticate the Tropea onions. A group from Italy used mass spectrometry to determine the concentration of 25 different elements in samples from the Tropea region as well as samples grown in nearby regions that are not included in the PGI designation.[3] With these elements they were able to build a multi-element profile that could then be statistically analysed to determine which elements were most useful in differentiating the origins of the onions. As it turned out, lanthanides, which are metallic chemical elements commonly referred to as rare earth metals, alkali metals (such as rubidium) and alkaline earth metals (such as strontium and calcium) were the most important elements in distinguishing the geographical origins of these onions. Dysprosium is a rare earth element that is particularly helpful in differentiating the origins of these onions, though it hadn't been considered in previous onion authentication studies. Onions are particularly efficient in taking up metals from soil, so it is not surprising that the metallic elements feature so prominently in distinguishing what soils they come from.

Faking the country of origin for a foodstuff may not simply be about gaining a premium price for specialised items from very specific regions, such as Tropea onions. More and more consumers are concerned about where their food comes from because they want to support local farmers, they want to minimise the miles their food has travelled or they want to avoid food that has come from regions known to have weaker production standards, maybe even a history of food fraud. Lying about where the food comes from helps to move it in a market that otherwise may not have bought it. For example, in 2014, executives of a family-based vegetable grower in Ontario, Canada, were caught mislabelling produce. They were charged with selling more than CA\$1,000,000 (£493,000) worth of vegetables that were not grown on their 400 acres (160 hectares) of greenhouse vegetable-producing land, nor anywhere else in Canada as they were labelled. They were mostly Mexican vegetables. They sold the produce on to major Canadian supermarkets where customers no doubt felt good about buying a Canadian product. The case went to pre-trial stage in January 2015 and hasn't yet concluded at the time of writing this book.

European regulations to protect the reputation of products from certain geographical origins are a very helpful framework for protecting consumers. The Tropea onions have PGI designation, as we mentioned. Other products, such as prosciutto di Parma from Italy and many wines, have protected designation of origin (PDO), and other foodstuffs have traditional specialities guaranteed (TSG) designation, such as Italian mozzarella. As of 2008, there are 172 fruits, vegetables and cereals that have either PDO or PGI designation, which is about 4 per cent of all products with these origin designations. Some of these include strawberries from the French city of Nîmes, forced rhubarb from Yorkshire, England and chestnuts from Terra Fria, Portugal. The system is not dissimilar to the rules of the old guilds in that the designation carries a particular reputation; the producers will be keen to guard that reputation and have the legal framework to do so.

Before it gets to the consumer, there are advantages in falsifying the country of origin to get around certain import taxes or import bans (as we saw with fish). It can help avoid food safety testing that is conducted more rigorously on imports from certain countries as well. In 2012, EU member states, plus Norway and Iceland, collected 6,472 food samples from countries that are subject to stricter import controls. Of these, 7.5 per cent of the samples exceeded the legal limit for one or more pesticide residues, compared with 1.4 per cent of samples originating from European countries. Within certain countries, the percentage of samples exceeding maximum residue levels (MRLs) was quite high. Over 38 per cent of food samples originating from Malaysia exceeded MRLs; Cambodia, Vietnam and Kenya all exceeded 20 per cent; and India and China both had just under 20 per cent of their samples exceeding MRLs. Products that most frequently exceeded the legal limits included basil (44.3 per cent of samples), okra (27 per cent), grapefruit (17.9 per cent) and celery leaves (17.3 per cent).[4]

Not only might the pesticide residues exceed legal limits, but in some countries residues of banned pesticides are making an appearance in food. More than 40 per cent of foods tested in Bangladesh contain banned pesticides. Random testing in India in 2013 found banned pesticides in food, often in quantities more than a thousand times the legal limit. Vegetables like cabbage and cauliflower are being dipped in pesticides to keep them fresh while apples and oranges are coated in wax made with the chemicals to extend their shelf life. Levels of the organochlorine heptachlor were 860 per cent above the legal limits for aubergines (eggplants) in India. Rice tested had levels of the insecticide chlorfenvinphos, which is banned in the EU and US because of its toxic effects, 1,324 per cent above the legal limit. Food from countries with less rigorous pesticide regulations is subject to increased monitoring and if the country of origin is false it may escape testing. This is where this kind of fraud starts to become a human health concern.

There are different analytical methods for authenticating food origins and we have seen some of these already. The Tropea onions used trace and rare earth element analysis, which can be done using a number of different techniques. With honey and meat, multi-isotope ratio analysis was conducted and compared with known isoscapes. Combining elemental and isotope analyses seems to provide the most impressive results in determining food origins and as more work is done in this field, traceability of our foodstuffs will become more reliable.

Production perjuries

The way fraudsters stand to make the biggest financial gain on fruits and vegetables is by falsely labelling products as organically produced. Consumers are willing to pay up to twice the price for organic produce. The market demand for organically produced food in 2011 was estimated at £40 billion (US$62.8 billion), triple its 2002 value. The US is the biggest consumer of organic food.

So far, most scandals regarding the mislabelling of organically produced food have revolved around poultry – both the meat and the eggs. These cases have generally been brought to light as a result of whistleblowers rather than any form of routine analysis. Cases regarding mislabelled fruit and vegetables, however, are hard to find in the literature. Whether this is because they don't exist or because authorities are looking in the wrong places is, as yet, unclear. However, it hasn't stopped scientists from finding ways of authenticating organically produced food.

Metabolomics has been used to differentiate some organic ingredients from non-organic as there has been some evidence that the different cultivation methods affect the quantities of metabolites, such as antioxidants, in the plants. This approach was used to distinguish ketchup made from organically grown tomatoes from conventional ketchup, for example.[5] The researchers from Spain used different techniques in mass spectrometry to separate the different metabolite compounds in the two types of ketchup. They found that antioxidants, such as caffeic acid, were found in significantly higher levels in the organic ketchup compared with the conventional. In fact, overall organic ketchup had higher concentrations of phenolic compounds, which are secondary metabolites in plants that are generally associated with higher nutritional value. However, the utility of this method over the long term is questionable because of that ever-present challenge of natural variability. These chemical signatures can evolve between different production years and depending on the varieties and origins of the tomatoes used.

Perhaps targeting differences in the production methods themselves is more effective. Each country and indeed each organic certification label has its own standards and criteria for organic production. In the EU, one of the requirements of organic farming is that no synthetic pesticides or synthetic mineral fertilisers can be used. In terms of analysis, this is a good starting point for differentiating organically produced fruit and veg from conventional production.

In 2007, scientists from the FSA and University of East Anglia (England) used nitrogen isotopes to try and differentiate organically grown tomatoes, lettuces and carrots from those that were conventionally grown.[6] The nitrogen in synthetic fertilisers comes from atmospheric nitrogen and as little fractionation happens in the manufacturing of these fertilisers, the stable isotope ratio of the fertilisers is very close to that of atmospheric nitrogen. Fertilisers used in organic production, however, have usually been through the food chain. As the ratio of ^{15}N to ^{14}N tends to increase with each trophic level, this means that the nitrogen isotope ratio of organic fertilisers is much higher than that of atmospheric nitrogen. When the researchers tested tomatoes, they found that although organically grown tomatoes did have a higher nitrogen isotope ratio, there was enough overlap between the lower values of organically grown tomatoes and the higher values of conventionally grown tomatoes to be ambiguous. There was even more overlap with lettuces and there was almost complete overlap when it came to the carrots. Carrots have a much lower nitrogen requirement than either tomatoes

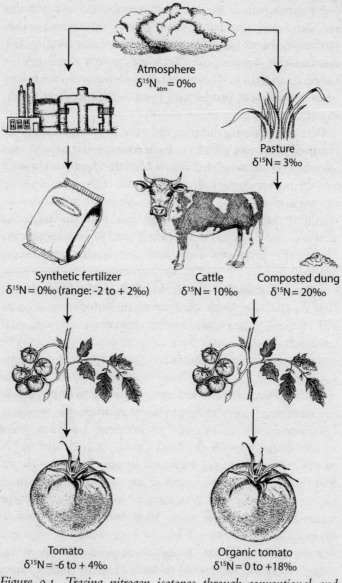

Figure 9.1. Tracing nitrogen isotopes through conventional and organic farming practices.

or lettuces and are more often cultivated in open fields rather than polytunnels or greenhouses, which may explain this difference. The conclusion was that nitrogen isotope ratios may provide supporting evidence in an authenticity case, but could not be considered conclusive evidence and were limited in their utility depending on the crop type. A different test was needed.

In 2014, German scientists tested the suitability of using ^{1}H NMR to build chemical profiles for organic and conventionally grown tomatoes.[7] The spectra showed significant differences between the two production methods for two cultivars of tomato. The results are promising, but limited. As we've mentioned previously, having a database of chemical signatures of different varieties of tomato cultivated using different farming practices and from different regions of the world is critical to understanding the natural variability – the NMR landscape for tomatoes, if you will. Further studies will have to be conducted as well to determine whether the analysis can be applied to other fruits and vegetables.

For now, it would appear that the authenticity of organic food still very much relies on the enforcement of strict production standards through rigorous certification processes and site inspections. In other words, a lot of paperwork.

Pulses, grains and seedy transactions

We decided to add pulses, grains and cereals to this chapter because, quite frankly, they didn't fit anywhere else and there is fraud happening in even these most basic of foodstuffs.

Rice appears on the food fraud databases as both an adulterated food and an adulterant of food. Ground into powder, rice has been used to adulterate spices, milk, wine and meat products. However, it has had its fair share of being on the receiving end of fraud as well. In 2011 headlines in Asia claimed that fake rice was being mass-produced in China. The rice was a mixture of potato starch and plastic resin. However, either the authorities covered up the scandal well or it could never be verified because it didn't go beyond a few headlines. Most rice fraud, however, has to do with the mislabelling or substitution of cheaper varieties as more expensive varieties, mainly basmati. In 2014, a joint effort by Interpol and Europol resulted in the seizure of 1,200 tonnes of counterfeit food, which included 22 tonnes of standard long-grain rice that was claiming to be basmati. Basmati can sell for two to three times the price of other long-grain rice. As well as this premium price on the shelves, brown basmati has no import duty in the UK, whereas other types of rice carry import costs of around £105 (US$164) per tonne. The UK will tolerate as much as 7 per cent non-basmati rice in a bag labelled as basmati, but beyond that it's considered intentional and fraudulent. DNA markers are used to detect and quantify adulteration of basmati.

In India, pulses (peas and lentils) have been found dyed with synthetic colours and toxic compounds such as lead chromate and metanil yellow. These dyes can be detected using chemical methods that Accum would have used in his time: adding hydrochloric acid will indicate the presence of both lead chromate and metanil yellow. In addition to pulses being dyed, seeds from common vetch (*Vicia sativa* L.) have been substituted for the split red lentil (*Lens culinaris* L.). HPLC has been used to look for chemical markers in vetch that aren't found in the red lentil, β-cyanoalanine and γ-glutamyl-β-cyanoalanine, to detect this adulteration.

As with any other food product, grains, pulses and seeds are more prone to fraud when they are processed beyond anything other than the raw commodity. A perfect example of this is grinding corn, chickpeas, wheat, soy, rice and other cereals and grains down into flour. As we mentioned in Chapter 6, the Chinese milk melamine scandal was preceded by the discovery of melamine in pet food in the US. Evidence suggests that the source of the melamine in the pet food was contaminated wheat gluten. After this scandal, the CFIA's Food Safety Division devised a list of odourless, colourless, tasteless and commercially available high-nitrogen compounds that could similarly be used to fraudulently increase the protein content of protein-containing foods, including a number of compounds used in fertilisers, such as urea. Scientists from the US FDA quickly devised methods to detect six compounds that they deemed most likely to be used as adulterants. They used a combination of chromatography (separation) and mass spectrometry (amount and type of chemical) techniques to detect the compounds in soy protein, wheat flour, wheat gluten and corn gluten. It's an example of the science of detection staying one step ahead of the fraudsters.

Of course, less sophisticated adulterants have also made their way into different flours, with the most popular being chalk powder. Substitutions of cheaper flours for more expensive flours also happen – wheat for spelt, for example. There are potential health consequences of such substitutions as people with intolerances, allergies or coeliac disease may be eating wheat flour inadvertently.

Rotten apples

This is a shorter chapter, and quite frankly we're happy about that. There are few horrific stories out there about our fruits and vegetables. OK, formalin-sprayed mangoes, plastic rice and pesticide-laced aubergines aren't great, but more than any other foodstuff we can control the fruits and vegetables we eat and reduce our vulnerability to food fraud. We repeat our mantra of 'buy whole and buy from who you know and trust'.

More alarming in this chapter, however, is what we're not allowed to see. How our basic fruits and vegetables are being processed, dipped in nanomaterials and other preservatives to fool us into thinking they are fresher than they are. As consumers we are not being given all the information about our food to enable us to make informed decisions. Is it assumed that we know there are processing methods out there that keep fresh food looking fresh well beyond what we know to be realistic? Or is it that they don't want us to know? Surely it has to be one or the other. As consumers, if we had that information, wouldn't we almost always choose the apple that rots?

CHAPTER TEN

Thoughts for Digestion

We challenge you right now to briefly put down this book and find yourself something small and full flavoured that you can pop into your mouth (but don't do it yet). A jellybean may not be the most nutritious option, but they work particularly well for this exercise. Now, when you've finished reading these instructions, put the book down, hold your nose and pop the jellybean into your mouth. KEEP HOLDING YOUR NOSE! Give the jellybean a good chew, swirl it around in your mouth and think about what flavours you are perceiving and then, while still chewing, let go of your nose and keep chewing for a bit longer. OK, now that you've read this, put down the book and do the exercise. We'll wait for you ...

Welcome back. Hopefully, unless you're anosmic (unable to smell), what you've just experienced is one of the most sophisticated pieces of equipment out there for conducting chemical analyses: the human sensory system. While we have stated over and over again throughout this book that most forms of food fraud are not detectable to the consumer, we would be remiss in completely ignoring this particularly economical and convenient piece of equipment that is, quite literally, right under our noses.

The human detector

The great science essayist, Lewis Thomas, wrote about smell: 'To be sure, I know that odor of cinnamon or juniper and can name such things with accuracy when they turn up in front of my nose, but I cannot imagine them into existence.'[1]

Most of us find it easy to bring up an image of a bicycle in our mind or recall the sound of a bird call with enough accuracy to replicate it. And though we may be able to label a smell accurately when we detect it or even describe a smell with reasonable accuracy, we generally find it very difficult to recall a smell 'into existence' as Thomas so eloquently puts it. Yet olfactory cues are rather critical to our existence. On the most basic level they convey information that helps us make decisions that affect our survival – such as smelling smoke or whether a substance is toxic. Our nasal passages play a role in helping us choose the right sexual partner, avoiding sickness and, as is relevant to this book, selecting food. Beyond that, smells enrich our lives in ways that we probably can't appreciate until the sense itself is removed.

Smells are made up of combinations of chemical odorants – molecules that evaporate and become airborne. The human olfactory system can detect and identify thousands of these odorants, and owing to our genetic makeup, cultural upbringing and personal experiences, two people can react very differently to exactly the same odorants.

When we breathe in, we draw chemical odorants into our nasal passages. These passages are lined with a thin sheet of mucus-coated sensory tissue, known as the olfactory epithelium. The odorant molecules get trapped in the mucus and make contact with olfactory receptor cells, which are nerve cells with a direct connection to the brain. The human nose contains hundreds of different types of olfactory receptors, whereas dogs may have over a thousand different types.

When the odorant molecule binds to one or more receptor cells, a biochemical chain reaction is triggered within the cell that causes a series of electrical pulses to be sent along the nerve fibres – known as axons – to the brain. The millions of axons from all of the nasal receptor cells are bundled together to form the olfactory nerve (the nose equivalent of the eye's optic nerve).

This electrical signal, carried along the olfactory nerve, is sent to the olfactory bulb, which is a concentration of nerves located in the forebrain just behind and between the eyes. This is where the brain begins to process the information. The nerve endings of the receptor cells cluster together in regions known as glomeruli, which act as switchboards of the olfactory system. As the electrical signals come in from various receptors, this switchboard 'lights up' in unique patterns that are then processed by the olfactory bulb.

The olfactory bulb then passes this information on to the rest of the brain where it is processed further. The information goes to the limbic system – a part of the brain involved with emotion and memory – as well as the olfactory cortex and orbitofrontal cortex. It is this additional processing that scientists think is essential in forming lifelong memories, complete with emotion, which are connected with smells.

It's commonly thought that humans have a relatively poor sense of smell compared with our fellow mammals. Evolution, after all, has reduced the size of our snouts as well as the number of genes that code for olfactory receptors. Rodents, for example, have 1,100 functional genes that code for olfactory receptors, whereas we humans have a mere 350. Does this mean that we have less capacity to smell our environment than the common sewer rat?

Not necessarily. Behavioural studies that test smell perception in humans and other primates suggest that we do as well as or better than other mammals.[2] In fact, we can even outperform our canine friends and the most sensitive measuring instruments when sniffing out particular odours.[3] Our 350 smell receptors are capable of detecting thousands of different odours, some with such sensitivity that we could detect less than a drop in an Olympic-sized swimming pool. One of the aromatic compounds in citrus for example, (Z)-8-tetradecenal, which, as you would have guessed, has a fruity, citrus-like odour, can be detected by humans at a threshold of 0.009 parts per billion when in water.[4] So although we have fewer olfactory receptors than a common sewer rat, our post-processing of this information in our brains somehow compensates and gives us a diverse and sensitive understanding of odours.

There are numerous selective advantages to having a good sniffer. Bacteria and viruses – the root cause of many illnesses – can change the body's chemistry directly or indirectly by eliciting immune responses. This is metabolomics at its core. These chemical changes might happen before a sick person becomes symptomatic. Perhaps their sweat or breath starts to smell different before they become feverish, for example. An ability to detect these changes in body chemistry at a time when the sick person is contagious but asymptomatic would be incredibly advantageous in avoiding illness.

There is evidence (scientific and anecdotal) to support this. Melanoma cells (cancerous skin cells), for example, produce compounds that aren't detected in normal melanocytes (skin cells). These compounds – dimethyl disulphide and trisulphide – are released as part of the vapour signature associated with a melanoma cell, giving them a distinct smell compared with normal cells. In 2014, scientists from Sweden and the US infected healthy individuals with an endotoxin, large molecules found in some bacteria that trigger a strong immune response in animals.[5] They found that within hours of being injected, individuals had a more unpleasing body odour relative to when they were exposed to a placebo. It was the first experimental evidence that an activated immune response is smelly, and that this can be detected by other humans so that they might avoid personal contact with affected individuals. Of course our use of deodorants, antiperspirants, perfumes and colognes mask all of these subtle cues these days.

As the jellybean exercise should have highlighted, our olfactory system is a crucial component of our ability to perceive flavours. The other major system, of course, is our gustatory system – our taste receptors. While our nose is the gatekeeper to determine whether to put something in our mouth, our sense of taste is what helps us decide whether to swallow it. The latest science suggests that humans can detect five groups of taste qualities: sweet, sour, bitter, salty and savoury (also known as umami). Not all animals can detect all these tastes – cats, for example, have been obligatory carnivores for so long in their evolutionary history that they have lost the ability to detect sweet flavours.

Each of the five taste qualities is stimulated by specific chemicals in the food. These chemicals are dissolved in saliva and then recognised by receptor cells in the tips of taste buds located in the mouth. Humans have about 10,000 taste buds, which are mostly clustered in small rounded bumps, called papillae, found on the surface of the tongue, though there are taste buds on the roof of the mouth, epiglottis and throat too.

When a taste receptor recognises a chemical stimulus – glutamate, for example, which is responsible for the broth-like umami taste quality – a series of biochemical reactions are triggered. Enzymes and hormones are released in the mouth, but the digestive system is also warmed up, knowing food is on its way. The receptor will also communicate with other cells within the taste bud, which translate the chemical information into an electrical message, much like olfaction. The electrical signal is carried along nerves to an area of the brain called the nucleus of the solitary tract (NST). Initial taste processing is done here and then signals are passed along to higher brain centres. It is there that this information is combined with smell information as well as chemical irritation, texture and appearance to define a comprehensive concept of flavour associated with the food, and then this information gets stored in our food reference database.

So surely this sophisticated sensory system is capable of detecting a little food fraud? We are, after all, capable of outperforming some chemical analyses. Where there are no varietal analyses for wines there are sommeliers who can, with repeatable accuracy, name the varietal as well as other descriptors of the wine. If you recall from Chapter 3, chemical analyses gave extra virgin olive oils a pass grade, yet the same ones failed the human sensory evaluations. Sensory panels are fine-tuned like any other analytical instrument. Individuals are given threshold tests to determine what their sensitivities are to certain chemical compounds. Then, with this information to hand, a panel can be carefully chosen that may be particularly sensitive to certain chemical compounds that are being sought out in the analysis, or alternatively, to make the panel as diverse as possible. The individuals are all trained to describe the flavour sensations they perceive objectively. They may taste hundreds of samples and have some familiarity with what compounds can create certain sensory experiences. They play an important role and will probably continue to do so in terms of food analysis. But what about those of us who have less finely tuned palates?

There is no denying that the average consumer is capable of detecting some forms of food fraud, not just based on smell and taste, but by combining that internal reference database we mentioned in Chapter 2 with all of our sensory inputs. When we open an envelope of saffron, does it smell like saffron? Does it look and behave like saffron? Does the fish fillet fall apart unexpectedly while cooking? Does the egg have a shell membrane when you crack it? And yet, people are being fooled daily by seemingly clumsy frauds.

With the exception of those that are anosmic, the downfall in the human fraud detection system does not lie in the equipment but in the reference database. The sensory system is there, but the database is sparse, or in some cases warped. As we mentioned in Chapter 1, a child that has been raised on foods with a synthesised strawberry flavour will still be able to tell a real strawberry from a fake one, but she will have learned a preference for the fake flavour. In her database, the compound ethyl methylphenylglycidate has been labelled as 'strawberry' while the cocktail of compounds associated with a real strawberry may be labelled as 'strawberry (other)'. People who have only ever eaten commercially produced loaves of bread will have 'seven to ten days' listed under the 'shelf life' category in their database, while those who have baked their own bread will have 'two to three days'. So if our databases for unadulterated authentic food are skewed, what chance do we have of recognising the fakes? If our brain is compensating for our olfactory receptor shortcomings, surely we must equip it with the information necessary for processing.

Technology to the rescue?

While we have far more superior noses than we give ourselves credit for, most food fraud is realistically beyond our sensory capabilities. With so many sophisticated analytical tests and so much technology available, is there any chance that anti-food-fraud devices will become available for consumers to conduct their own testing?

Perhaps we just need an unbiased electronic version of the equipment we already have. The quest for electronic noses and tongues has been going on for decades. Their allure is that they have the potential to be relatively cheap, relatively mobile and easier to use than other methods that examine volatile compounds. But as we described in Chapter 3, there have been challenges in achieving the same sensitivity and specificity as a human sensory panel can provide. Versions of these devices, particularly the electronic or e-nose, have found applications in the food and beverage industry. Most of these applications revolve around spoilage detection, which is likely because most human noses would probably rather avoid such smells. Fish spoilage, for example, has been thoroughly investigated using electronic noses – the volatiles produced as the flesh deteriorates are measured by the e-nose and compared with other observable changes such as colour change.

Researchers from Barcelona, Spain, have been developing an electronic tongue, which they first used to help discriminate between different types of cava wines and, more recently, types of commercial beer. They hope that eventually they will get it to a point where it can be used to help detect fraud – but, once again, it will depend on the type of fraud. Such devices will no doubt find applications within the food and beverage industry, but it's unlikely that there would ever be a pocket-sized version for consumers. After all, how could it ever be cheaper and more convenient to use than our own nose and tongue?

Dreams of hand-held consumer testing devices stumble upon the same obstacles that the food analyst suffers when trying to battle food fraud: there's a different test for every question. So while formaldehyde indicator strips can be purchased online and fit easily in a pocket or handbag, they only answer one question. The technology exists already to build a pregnancy-kit-like device to determine whether your swordfish is really swordfish, but again, it's a single-use, single-question test. For now, the concept of a *Star Trek*-style tricorder that is a multifunctional hand-held device for detecting food fraud remains in the realm of science fiction, though perhaps advances towards such a device in the medical field might speed things along.

What we do see coming down the pipeline are new modes of delivering information about traceability. This is known as the food traceability tech sector and it's estimated that the world market for such technologies will reach over £7 million (US$ 11 million) in 2015. Simply put, these are technologies that connect consumers with their food's journey. The seafood industry seems to be working particularly hard at developing these types of technologies. We mentioned ThisFish and some other projects in

Chapter 4, but there are others as well, many of which will be rolling out phone apps about the same time that this book is released. The hard work in these systems is building the database to support them. The technology is just a matter of a QR code and a smart phone – it's tracking down the supply chain and building the story behind the food that's the hard part.

One Degree Organic Foods in Abbotsford, Canada has developed such a system for their products. A shopper can scan a QR code on their product label or plug a six-digit product code into the One Degree website and trace every ingredient in that product back to its source. There are even videos of the farmers to give the customer a greater sense of their food and where it came from. It's one of the first places to offer this to multi-ingredient products. What this means for the business, however, is that nearly half the business is devoted to the technology side of things.

The logical step for companies not willing or capable of devoting such resources to developing these types of technologies is to bring in third-party authenticity certification schemes. Robert Hanner and his group at the University of Guelph created a spin-out company, called TRU-ID, which uses DNA barcoding technology to authenticate food and natural health products. As consumers put greater value on the authenticity of their food, producers may gain a competitive advantage in having such certifications identified with their products. On the other hand, there is also the potential for abusing the certification, as we have seen with organic certifications.

It would seem, for now at least, that technology is not going to equip us with analytical tools that we can take with us to the supermarket. However, we should expect to see more of it in the very near future helping us get better connected to our food. Even as we write this last sentence, something seems wrong. We need technology to connect us with the stuff we grow in the ground? On one hand we think it's wonderful to be better connected to our food and have tools to explore where it has come from, if for no other reason than that it forces retailers to audit their own supply chains. On the other hand, it feels not dissimilar to standing on a corner staring at your phone to find directions when you could simply ask someone. There are immeasurable and unexpected benefits of just talking to people.

Taking action

While we await the pocket version of the electronic tongue, there are other actions we can take as consumers to try and minimise our exposure to food fraud. In each of the chapters we have tried to provide advice specific to that type of food, but it isn't always practical to apply it more generally. For example, buying fish with its head on is a good idea, but this doesn't really work with other meat products (though there might be more vegetarians in the world if it did)!

Before we go further, however, let us just say that we are fully aware that buying food can be complicated. We may enter a shop with our moral compass directing us towards fresh, whole, organic, local and ethically sourced food. And then screaming children, loud-mobile-phone-talker guy, chatting ladies who won't get out of the way, disgruntled I-don't-get-paid-enough-to-know-what-chia-seeds-are employees, and what seems like a never-ending shelf restocking brigade all get in the way (sometimes quite literally). The next thing you know, you're checking out with a bunch of ready-made meals, comfort food (aka chocolate and crisps) and highly packaged lunch-box items. None of it was on your grocery list and none of it resembles real food. The moral compass got lost somewhere in the frozen foods section and you feel guilty because you feel you have somehow let yourself, your family, independent grocers, animals, local farmers and the rest of the planet down because you made some poor purchasing decisions. And now you have to consider food fraud as well?

Relax. There are a lot of decisions to make around the food that we buy – price, nutritional value, taste, what your household will eat, ethics, sustainability and yes ... authenticity. Our priorities will change depending on our financial situation in that moment, who's coming for dinner that week, how busy we are, what food scandal has hit the media lately, how hungry we are and what mood we're in. That's just life.

But information is a powerful tool and the more informed we are, the better decisions we can make. Estimates are that one in every ten items we buy in the supermarket is fraudulent in some way, so we need to go in there armed with whatever we can. So, in the interest of adding to your toolbox, here are some actions that may help reduce your vulnerability to food fraud.

Buy whole, recognisable foods. This isn't always practical of course, but it is much harder to fake a whole almond than it is to add an adulterant to ground almonds. Fish fillets that have no recognisable features can easily be mislabelled as other species. Eat fruit whole or juice it yourself. Whole spices are not only less prone to fraud, they keep longer in the cupboard. This is probably the single most effective action we as consumers can take towards reducing our vulnerability to fraud and probably improving our diet at the same time.

Rein in the chain. If you buy whole recognisable foods, the food supply chain will naturally shorten. However, there are other ways to do this. Honey was the perfect example of this – find out where your local beekeeper is and buy direct. If you're lucky enough to live near a farm shop or farmer's market, take advantage of this as often as you can. If you don't have a market, look into schemes that may deliver locally produced food in a weekly box. And don't be afraid to ask retailers who they buy from or more generally what their supply chains look like.

Buy from people you trust. This doesn't necessarily mean that you have to play poker every Friday with your local fishmonger. But it does mean that you should feel confident that if something was really wrong with your food – for instance, if it made your family sick – the person or retailer you bought it from would be able to provide you with quick answers about the product they sold you. Somewhere in our research we read that you can learn far more from the person selling you the food than you can from any label – sage advice indeed.

Don't fall prey to unrealistic prices. We've seen over and over again that prices that seem too good to be true probably are. If you come across such deals, ask yourself how they're possible. If you're feeling particularly brazen, ask the retailer how such a deal can exist. How can a bottle of vodka be worth two quid? Food (not that we really consider vodka food) should not be cheap – be realistic about the cost of food.

Expand your internal food reference database. We all need to refresh (in some cases reboot completely) and expand our internal food reference databases. Learn how it is made but also go out into the world and explore food or take it home and experiment with it. This means experiencing our food. In this hectic world it is all too easy to wolf down some form of nutrition while on the move or at our desks. Mindless eating has to be one of the downfalls of our modern society. If we consider each meal as an opportunity to expand our database, perhaps we will take more time to understand and appreciate our food.

This is also where education plays a role. Chef Jamie Oliver is leading the charge in this area with the campaign Food Revolution Day. The goal is to make practical food education compulsory in the school curriculum, globally (nobody could ever call Oliver an underachiever). The motivation behind the campaign is to help children lead healthier and happier lives, while tackling the global obesity problem. However, there are potential side effects to this in that educating children about good food starts to build their reference database. Perhaps it even means that ethyl methylphenylglycidate gets reassigned with the label 'fake strawberry'.

Find the story in your food. The criteria for choosing our food can sometimes seem overwhelming. Often this can be simplified by selecting food that has a good story. Let us give you an example. One story is: I bought some mature cheese from the supermarket. It wasn't particularly mature-tasting, but it was on sale. A second story is: I popped into this cheese shop and started talking to the guy behind the counter (we promise this isn't the Monty Python sketch all over again). He's been making and selling cheese for 30 years. He took the business over from his mother when his older sister didn't want it. He let me try a whole bunch of cheeses, but I told him I was really after something very sharp. You know, the kind of cheese that takes the skin off the roof of your mouth. So, he took me into the back where he ages his own cheese and cut off a piece of this round he had ... well, it was the best cheese I've ever had. Bought the whole round off the guy!

OK, so these are obviously two extremes, but you get the picture. We don't realise how much we want to know the story behind our food until we get it. This is perhaps one of the reasons why Riverford Organics has been so successful in the UK. Each week, along with a box of food, there is a story about what's happening at Riverford. And it's not some gloss-over of what's in bloom or how great the carrots are looking, though there's some of that as well, obviously. It's quite often the bad and the ugly stories, from the fields as well as the office. Details like 'ten weeks of solid rain have meant that planting's delayed so expect your lettuce to be a little later this season'. In the supermarket, if the lettuce isn't on the shelf it's because some truck didn't show up. Disasters and frustrations are inherent in farming – producing food isn't always easy. As we begin to find the story of our food, other things slip into place at the same time. We are attracted to the foods with shorter supply chains and we develop relationships with the people who provide it. We are willing to pay a fair price for it because we understand what went into making it. And we start to update our own reference database under this new paradigm. It's unrealistic to think that all of our food will have a story, but we tend to appreciate it more when it does.

Final words

Whether motivated by Horsegate or some other food scandal, governments around the world seem to be moving food fraud further up the priority list. This is reassuring, but we must also be realistic. Governments aren't likely to show great concern over a little unlabelled added water in our ham when three-quarters of supermarket chickens are contaminated with campylobacter or when people are dying from an $E.$ $coli$ O157:H7 outbreak. But, having fallen victim to some tasteless (pun intended) practical jokes, governments are looking at how to avoid being so gullible next time and reducing vulnerabilities in the food supply. As some food fraud scandals have shown us, there can be serious health consequences to some types of fraud. These have been acute cases; we have yet to learn what chronic implications there may be.

As food authenticity becomes a greater factor in consumer choice, the food industry will also be further motivated to audit its processes and reduce its exposure to food fraud. As consumers, our role is to hold the industry accountable.

We must be prepared, however, for food fraud to become more prevalent before it gets better. Climate change combined with rising demand creates an ideal environment for fraud. Intelligence gathering, such as is being done by the NCFPD (described in Chapter 1), will help identify where in the food fraud haystack analysts should be looking: for example, flagging Spanish olive oil after a particularly poor yield of olives in Spain. Intelligence and technologies that enable food enforcement agencies and analysts to be proactive certainly won't put an end to food fraud, but they might keep them one step ahead in the scientific arms race against the fraudsters.

One of our greatest realisations in the process of writing this book is that food fraud is a continuum on many levels. There are actions that are clearly illegal (fake milk) and actions that are legal with an air of dishonesty (some fruit preservation methods). Food scandals range from harmless (billionaires battling over rare wine) to deadly (methyl alcohol in cheaper wine), and from financially insignificant (pennies-worth of added water) to market-crashing big business (the Soybean or Salad Oil Scandal of the 1960s). Even the mindset of the criminals who carry out fraud lies on a continuum from immoral criminal to creative problem-solver. Between these extremes lies everything in between. This world of food production, let alone food fraud, is far from black and white.

In order to be clearer on what constitutes food fraud we first need to resolve some of these rather grey areas that lie within the legal framework of food production. How are we being misled about our food with behind-the-scenes processing methods, blurring the lines of reality? How much of our food is being manufactured to pass certain tests? For example, if food is manufactured to meet nitrogen content requirements with no regard to the source of that nitrogen, is this not misleading in terms of its nutritive properties? Throw into this mix technologies that are developing faster than we can understand them, such as the use of nanoparticles, and we have an interesting future ahead of us.

We live within a complex food culture, it's true. But we are not powerless within it by any means. Just as we take action to minimise risk in other aspects of our lives – looking before we cross the street or protecting ourselves against sun exposure – we can be active in our food choices. As we have learned, many of the food forensic analyses are based on the concept of 'we are what we eat'. Humans are no different. The chemicals from our food become part of us and we should not be passive about what we are putting into our bodies. It is fundamental to our well-being. Our food system is such that many things lie outside our control, but there are always choices. We are, after all, the consumers, and only we decide what to put in our mouths. Choose well.

Appendix: Some of the Chemical Structures Discussed in the Book

Complex biochemical compounds (generally called 'organic compounds' amongst chemistry aficionados) are mainly made of the elements carbon (C), hydrogen (H), oxygen (O), nitrogen (N) and, to a lesser extent, sulphur (S) and phosphorous (P). The myriad of structures that are produced in nature and by the chemical industry arise because C can form four bonds with other elements. In particular, it can bond to itself to form linear chains, chains with branches, and rings (commonly hexagons and pentagons). In theory, an infinite number of structures is possible, but the biochemical pathways found in nature mean that the structures that occur are organised around commonly occurring themes. Examples of some of these can be seen below and in the text.

It is important to note that because C and H are such common elements we tend not to display most of these in the structures, otherwise even quite simple compounds would become hugely complicated. For example, a simple fatty acid looks like this when all the C and H atoms are shown.

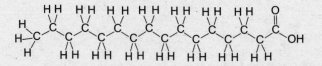

The shorthand form shown below, leaving out most of the letters for the C and H atoms, is the convention used throughout chemistry and biochemistry when we draw molecular structures.

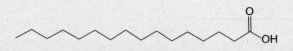

Chapter 1

Trichloroanisole (TCA) a natural compound that can give wines a musty or corky flavour

Chapter 3

Dimethylpolysiloxane (E900) – an industrial polymer added to vegetable oils to stop foaming during frying

Aniline – the denaturant added to vegetable oil to make it fit for industrial use only

Anilide – formed when industrial vegetable oil is refined to remove aniline

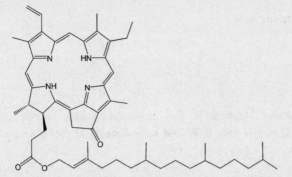

Pyropheophytin A (PPP) – a degradation product of chlorophyll, the compound that gives olive oil its green tinge

Chapter 4

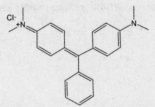

Malachite green – an effective treatment for fish diseases, but risks to human health mean it's banned for use in aquaculture in many countries

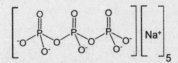

Sodium tripolyphosphate (TPP) – used to preserve seafood and, more dishonestly, in the underweighting of seafood

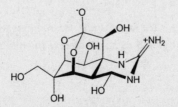

Tetrodotoxin – the toxic compound found in pufferfish

Chapter 6

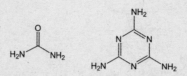

Urea and Melamine – Two industrial products high in nitrogen and added to fake milk and baby formula to fool simple protein tests

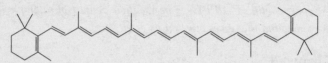

β-carotene – the main natural yellow-orange pigment occurring in cheese

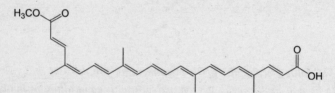

Bixin – the orange pigment found in annatto (E160b), used to colour cheese and cheese substitutes

Chapter 7

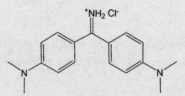

Auramine O – used as a stain for some types of bacteria in microbiology, but found as an adulterant in ground coriander

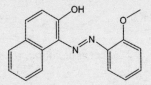

Sudan Red – Sudan dyes are used to colour waxes, oils, solvents and polishes, but have also been used to colour ground spices such as curry and chili powders

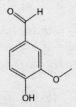

Vanillin – the compound behind the flavour and smell of natural vanilla, which is commercially synthesised to meet demand from the food and beverage industry

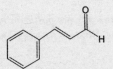

Cinnamaldehyde – the principal component of cinnamon

Chapter 8

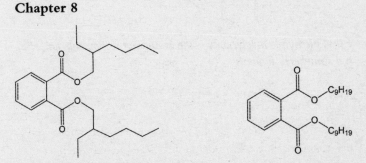

Di(2-ethylhexyl) phthalate and diisononyl phthalate – phtahlates used to increase the flexibility of plastics, such as PVC, but found being added to jams, sports drinks and fruit juices to make them appear more cloudy and therefore more natural

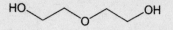

Diethylene glycol – a component of antifreeze, but illegally used to sweeten wine

Chapter 9

Formaldehyde – a naturally occurring compound with many industrial uses, but widely known as a tissue fixative and embalming agent, which is why it has been used to help preserve fruits and vegetables

Chapter 10

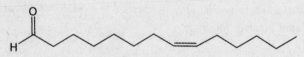

(Z)-8-tetradecenal – humans are particularly good at detecting this aromatic compound, which gives citrus its characteristic odour

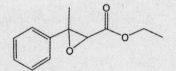

Ethyl methylphenylglycidate – the compound responsible for artificial strawberry flavours

Notes

Chapter 1

1   Food and Drug Administration. 2009. Economically Motivated Adulteration. Public Meeting; Request for Comment (Docket No. FDA-2009-N-0166) (Electronic Version). Federal Register, 74, 15497–9. http://edocket.access.gpo.gov/2009/pdf/E9--7843.pdf

2   Lewis, J. Lead Poisoning: A Historical Perspective. *United States Environmental Protection Agency* (EPA Journal, May 1985). http://www2.epa.gov/aboutepa/lead-poisoning-historical-perspective

3   Quoted in Wilson, B. 2008. *Swindled*. Princeton University Press, New Jersey, p. 163.

4   Ropicki, A., Larkin, S. & Adams, C. 2010. Seafood substitution and mislabeling: WTP for a locally-caught grouper labeling program in Florida. *Marine Resource Economics* 25: 77–92.

5   National Audit Office (10 Oct 2013). *Food safety and authenticity in the processed meat supply chain*. The Food Standards Agency, Department for Environment, Food & Rural Affairs, Department of Health, HC 685, Session 2013–2014: 10.

Chapter 2

1   Ng, C. M. & Reuter, W. M. 2015. *Application Note: Analysis of sugars in honey using the Perkin Elmer Altus HPLC system with RI detection*. http://www.perkinelmer.co.uk/CMSResources/Images/44-171789APP_Analysis-of-Sugars-in-Honey-012101_01.pdf

2   Di Girolamo, F., D'Amato, A. & Righetti, P. G. 2012. Assessment of the floral origin of honey via proteomic tools. *Journal of Proteomics* 75(12): 3688–93.

3   Schievano, E., Stocchero, M., Morelato, E., Facchin, C. & Mammi, S. 2012. An NMR-based metabolomic

approach to identify the botanical origin of honey. *Metabolomics* 8(4): 679–90.

4   Ohmenhaeuser, M., Monakhova, Y. B., Kuballa, T. & Lachenmeier, D. W. 2013. Qualitative and quantitative control of honeys using NMR spectroscopy and chemometrics. *ISRN Analytical Chemistry* 2013: 825318.

5   Beitlich, N., Koelling-Speer, I., Oelschlaegel, S. & Speer, K. 2014. Differentiation of manuka honey from kanuka honey and from jelly bush honey using HS-SPME-GC/MS and UHPLC-PDA-MS/MS. *Journal of Agricultural and Food Chemistry* 62: 6435–44.

6   Schnell, I. B., Fraser, M., Willerslev, E. & Gilbert, M. T. P. 2010. Characterisation of insect and plant origins using DNA extracted from small volumes of bee honey. *Arthropod-Plant Interactions* 4: 107–16.

7   Bruni, I., Galimberti, A., Caridi, L., Scaccabarozzi, D., De Mattia, F., Casiraghi, M. & Labra, M. 2015. A DNA barcoding approach to identify plant species in multiflower honey. *Food Chemistry* 170: 308–15.

8   Schellenberg, A., Chmielus, S., Schlicht, C., Camin, F., Perini, M., Bontempo, L., Heinrich, K., Kelly, S.D., Rossmann, A., Thomas, F., Jamin, E. & Horacek, M. 2010. Multielement stable isotope ratios (H, C, N, S) of honey from different European regions. *Food Chemistry* 121(3): 770–7.

Chapter 3

1   http://www.mcdonalds.com/us/en/your_questions/our_food/does-your-food-contain-dimethylpolysiloxane.html

2   Gelpí, E., Posada de la Paz, M., Terracini, B., Abaitua, I., Gómez de la Cámara, A., Kilbourne, E. M., Lahoz, C., Nemery, B., Philen, R. M., Soldevilla, L. & Tarkowski, S. (WHO/CISAT Scientific Committee for the Toxic Oil

Syndrome). 2002. The Spanish toxic oil syndrome 20 years after its onset: A multidisciplinary review of scientific knowledge. *Environmental Health Perspectives* 110(5): 457–64.

3   Posada de la Paz, M., Philen, R. M. & Abaitua Borda, I. 2001. Toxic oil syndrome: The perspective after 20 years. *Epidemiologic Reviews* 23(2): 231–46.

4   Woodbury, S. E., Evershed, R. P., Rossell, J. B., Griffiths, R. E. & Farnell, P. 1995. Detection of vegetable oil adulteration using gas chromatography combustion/isotope ratio mass spectrometry. *Analytical Chemistry* 67: 2685–90.

5   Woodbury, S. E., Evershed, R. P. & Rossell, J. B. 1998. Purity assessments of major vegetable oils based on $\delta^{13}C$ values of individual fatty acids. *Journal of the American Oil Chemists Society* 75(3): 371–9.

6   Mottram, H. R., Woodbury, S. E., Rossell, J. B. & Evershed, R. P. 2003. High-resolution detection of adulteration of maize oil using multi-component compound-specific $\delta^{13}C$ values of major and minor components and discriminant analysis. *Rapid Communication in Mass Spectrometry* 17: 706–12.

7   Frankel, E. N., Mailer, R. J., Wang, S. C., Shoemaker, C. F., Guinard, J.-X., Flynn, J. D. & Sturzenberger, N. D. 2011. *Report: Evaluation of Extra-Virgin Olive Oil Sold in California.* UC Davis Olive Centre.

8   European Commission, Research & Innovation, Funding Opportunities. Call: H2020-SFS-2014-2. https://ec.europa.eu/research/participants/portal/desktop/en/opportunities/h2020/topics/762-sfs-14a-2014.html

Chapter 4

1   Tennyson, J. M., Winters, K. S. & Powell, K. 1997. *A Fish by any Other Name: A Report on Species Substitution.* National Marine Fisheries Service, National Seafood Inspection Laboratory. MS 39568-1207.

2   Oehlenschläger, J. & Rehbein, H. 2009. Basic facts and figures. In Rehbein, H. and Oehlenschläger, J. (eds), *Fishery Products: Quality, safety and authenticity*. Blackwell Publishing, Vigo, Spain.

3   Rehbein, H. 2003. Identification of fish species by protein- and DNA-analysis. In R. I. Pérez-Martín & C. G. Sotelo (eds), *Authenticity of Species in Meat and Seafood Products*. International Congress on Authenticity of Species in Meat and Seafood Products.

4   Marko, P. B., Lee, S. C., Rice, A. M., Gramling, J. M., Fitzhenry, T. M., McAlister, J. S., Harper, G. R. & Moran, A. L. 2004. Mislabelling of a depleted reef fish. *Nature* 430: 309–10.

5   Hebert, P. D. N., Ratnasingham, S. & deWaard, J. R. 2003. Barcoding animal life: cytochrome *c* oxidase subunit 1 divergences among closely related species. *Proceedings of the Royal Society of London, B Series* 270: S96–S99.

6   Wong, E. H.-K. & Hanner, R. H. 2008. DNA barcoding detects market substitution in North American seafood. *Food Research International* 41: 828–37.

7   For a review of global seafood mislabelling, see Jacquet, J. L. & Pauly, D. 2008. Trade secrets: renaming and mislabeling of seafood. *Marine Policy* 32: 309–18.

8   Rasmussen Hellberg, R. S., Morrissey, M. T. & Hanner, R. H. 2010. A multiplex PCR method for the identification of commercially important salmon and trout species (*Oncorhynchus* and *Salmo*) in North America. *Journal of Food Science* 75: C595–C606.

9   Cohen, N. J., Deeds, J. R., Wong, E. S., Hanner, R. H., Yancy, H. F., White, K. D., Thompson, T. M., Wahl, M., Pham, T. D., Guichard, F. M., Huh, I., Austin, C., Dizikes, G. & Gerber, S. I. 2009. Public health response to puffer fish (tetrodotoxin) poisoning from mislabeled product. *Journal of Food Protection* 72(4): 810–17.

10   Quote extracted from: Rentz, C. (6 Dec 2014). Seafood
      fraud cases plummet as NOAA cuts investigators.
      Baltimore Sun.

Chapter 5

1   The Animal By-Products Regulations 2002. Regulatory
      Impact Assessment Annex A. http://www.food.gov.
      uk/sites/default/files/multimedia/pdfs/RIAanimal
      byproductsscot.pdf, pp. 5–6.
2   Quote taken from *BBC News* (9 Jun 2004). Good
      enough to eat? BBC. http://news.bbc.co.uk/1/hi/
      business/3087011.stm
3   Painter, J. A., Hoekstra, R. M., Ayers, T., Tauxe, R. V.,
      Braden, C. R., Angulo, F. J. & Griffin, P. M. 2013.
      Attribution of foodborne illnesses, hospitalizations,
      and deaths to food commodities by using outbreak
      Data, United States, 1998–2008. *Emerging Infectious
      Diseases* 19(3): 407–15.
4   Quote taken from Lawrence, F. (22 Oct 2013). Horsemeat
      scandal: where did the 29% horse in your Tesco burger
      come from? *Guardian.* http://www.theguardian.com/
      uk-news/2013/oct/22/horsemeat-scandal-guardian-
      investigation-public-secrecy
5   Hsieh, Y. H. P., Woodward, B. B. & Ho, S. H. 1995.
      Detection of species substitution in raw and cooked
      meats using immunoassays. *Journal of Food Protection*
      5: 555–9.
6   Ayaz, Y., Ayaz, N. D. & Erol, I. 2006. Detection of
      species in meat and meat products using enzyme-
      linked immunosorbent assay. *Journal of Muscle Foods*
      17(2): 214–20.
7   Greenfield, H. & Kosulwat, S. 1991. Nutrient
      composition of Australian fresh retail sausages and the
      effects of cooking on fat content. *Journal of the Science
      of Food and Agriculture* 57: 65–75.

8 Cawthorn, D.-M., Steinman, H. A. & Hoffman, L. C. 2013. A high incidence of species substitution and mislabelling detected in meat products sold in South Africa. *Food Control* 32(2): 440–9.

9 FSA. *Study into Injection Powders Used as Water Retaining Agents in Frozen Chicken Breast Products.* http://www. eurocarne.com/daal?a1=informes&a2=reportchicken-study.pdf

10 D'Amato, M. E., Alechine, E., Cloete, K. W., Davison, S. & Corach, D. 2013. Where is the game? Wild meat products authentication in South Africa: a case study. *Investigative Genetics* 4: 6.

11 Quote taken from Malthus, T. R. 1798. *An Essay on the Principle of Population.* Chapter 1. J. Johnson, in St. Paul's Church-yard, London.

Chapter 6

1 Evershed, R. P., Payne, S., Sherratt, A. G., Copley, M. S., Coolidge, J., Urem-Kotsu, D., Kotsakis, K., Özdoğan, M., Özdoğan, A., Nieuwenhuyse, O., Akkermans, P. M. M. G., Bailey, D., Andeescu, R.-R., Campbell, S., Farid, S., Hodder, I., Yalman, N., Özbaşaran, M., Biçakci, E., Garkinfel, Y., Levy, T. & Burton, M. M. 2008. Earliest date for milk use in the Near East and southeastern Europe linked to cattle herding. *Nature* 455: 528–31.

2 Vigne, J.-D. & Helmer, D. 2007. Was milk a 'secondary product' in the Old World Neolithisation process? Its role in the domestication of cattle sheep and goats. *Anthropozoologica* 42: 9–40.

3 Curry, A. 2013. Archaeology: The milk revolution. *Nature* 500: 20–2.

4 Salque, M., Bogucki, P. I., Pyzel, J., Sobkowiak-Tabaka, I., Grygiel, R., Szmyt, M. & Evershed, R. P. 2013. Earliest evidence for cheese making in the sixth millennium BC in northern Europe. *Nature* 493: 522–5.

5   Sansoucy, R. Livestock – a driving force for food security and sustainable development. FAO Corporate Document Repository. http://www.fao.org/docrep/v8180t/v8180t07.htm

6   Ulberth, F. 2003. Milk and dairy products. In *Food Authenticity and Traceability* (ed. Michele Lees). Chapter 16. Woodhead Publishing Ltd & CRC Press LCC, Cambridge and Boca Raton, Florida.

7   Mottram, H., Woodbury, S. E. & Evershed, R. P. 1997. Identification of triacylglycerol positional isomers present in vegetable oils by high performance liquid chromatography/atmospheric pressure chemical ionization mass spectrometry. *Rapid Communications in Mass Spectrometry* 11: 1240–52.

8   Mottram, H. R. & Evershed, R. P. 2001. Elucidation of the composition of bovine milk fat triacylglycerols using high-performance liquid chromatography–atmospheric pressure chemical ionisation mass spectrometry. *Journal of Chromatography A* 926: 239–53.

9   Mottram, H. R., Crossman, Z. M. & Evershed, R. P. 2001. Regiospecific characterisation of the triacylglycerols in animal fats using high performance liquid chromatography-atmospheric pressure chemical ionisation mass spectrometry. *Analyst* 126: 1018–24.

10 Bradley, H. W. Improvement in compounds for culinary use. US patent No. 110,626. January 3, 1871. http://www.google.com/patents/US110626

11 Raney, M. Method of producing finely-divided nickel. US patent No. 1,628,190. 10 May 1927. http://www.google.co.uk/patents/US1628190

12 Mozaffarian, D., Katan, M. B., Stampfer, M. J. & Willett, W. C. 2006. Trans fatty acids and cardiovascular disease. *The New England Journal of Medicine* 345: 1601–13.

13 Derewiaka, D., Sosińska, E., Obiedziński, M., Krogulec, A. & Czaplicki, S. 2011. Determination of the adulteration of butter. *Journal of Lipid Research and Metabolism* 113: 1005–11.

14   http://www.qsrmagazine.com/news/there-cheese-fraud-epidemic

15   http://www.bloomberg.com/news/articles/2014-08-26/swiss-combat-counterfeit-cheese-with-dna-fingerprinting

16   Branigan, T. (2 Dec 2008). Chinese figures show fivefold rise in babies sick from contaminated milk. *The Guardian*.   http://www.guardian.co.uk/world/2008/dec/02/china

17   Tyan, Y.-C., Yang, M.-H., Jong, S.-B., Wang, C.-K. & Shiea, J. 2009. Melamine contamination. *Analytical and Bioanalytical Chemistry* 395:729–35.

18   http://news.bbc.co.uk/1/hi/7720404.stm.

19   http://www.cfs.gov.hk/english/multimedia/multimedia_pub/multimedia_pub_fsf_56_03.html

Chapter 7

1   Curl, C. L. & Fenwick, G. R. 1983. On the determination of papaya seed adulteration of black pepper. *Food Chemistry* 12: 241–7.

2   Naseema, B. S., Ambily, P., Thomas, G., Biju, M. T., Pratheesh, K. N., George, X., Pradeep, K. G. T., Rajith, R., Prathibha, R. K. & Visal, K. S. 2014. Pesticide residues in soils under cardamom cultivation in Kerala, India. *Pesticide Research Journal* 26(1): 35–41.

3   Priyadarshini, S. 2014. Himalayas losing prized spice to climate change, poor science. *Nature India*: 10.1038/nindia.2014.162

Chapter 8

1   Jefferson's visit to Bordeaux is given in the history of Château Lafite: http://www.lafite.com/en/chateau-lafite-rothschild/history

2   Liu, L., Cozzolino, D., Cynkar, W. U., Gishen, M. & Colby, C. B. 2006. Geographic classification of Spanish and Australian tempranillo red wines by visible and

near-infrared spectroscopy combined with multivariate analysis. *Journal of Agricultural and Food Chemistry* 54(18): 6754–9.

3   Cozzolino, D., Smyth, H. E. & Gishen, M. 2003. Feasibility study on the use of visible and near-infrared spectroscopy together with chemometrics to discriminate between commercial white wines of different varietal origins. *Journal of Agricultural and Food Chemistry* 51(26): 7703–8.

4   Doner, L. W., Ajie, H. O., Sternberg, L. d S. L., Milburn, J. M., De Niro, M. J. & Hicks, K. B. 1987. Detecting sugar beet syrups in orange juice by D/H and $^{18}O/^{16}O$ analysis of sucrose. *Journal of Agricultural and Food Chemistry* 35: 610–12.

5   Le Gall, G., Puaud, M. & Colquhoun, I. J. 2001. Discrimination between orange juice and pulp wash by ^{1}H nuclear magnetic resonance spectroscopy: identification of marker compounds. *Journal of Agricultural and Food Chemistry* 49(2): 580–8.

6   Wu, M.-T., Wu, C.-F., Wu, J.-R., Chen, B.-H., Chen, E. K., Chao, M.-C., Liu, C.-K. & Ho, C.-K. 2012. The public health threat of phthalate-tainted foodstuffs in Taiwan: The policies the government implemented and the lessons we learned. *Environmental International* 44: 75–9.

Chapter 9

1   Zhang, Z., Kong, F., Vardhanabhuti, B., Mustapha, A. & Lin, M. 2012. Detection of engineered silver nanoparticle contamination in pears. *Journal of Agricultural and Food Chemistry* 60(43): 10762–7.

2   Spink, J. & Moyer, D. C. 2011. *Backgrounder: Defining the Public Health Threat of Food Fraud.*Anti-Counterfeiting and Product Protection Program, Michigan State University.

3   Furia, E., Naccarato, A., Sindona, G., Stabile, G. & Tagarelli, A. 2011. Multielement fingerprinting as

a tool in origin authentication of PGI food products: Tropea red onion. *Journal of Agricultural and Food Chemistry* 59: 8450–7.

4   European Food Safety Authority. 2014. The 2012 European Union Report on pesticide residues in food. *EFSA Journal*, 12(12): 3942.

5   Vallverdú-Queralt, A., Medina-Remón, A., Casals-Ribes, I., Amat, M. & Lamuela-Raventós, R. M. 2011. A metabolomic approach differentiates between conventional and organic ketchups. *Journal of Agricultural and Food Chemistry* 59: 11703–10.

6   Bateman, A. S., Kelly, S. D. & Woolfe, M. 2007. Nitrogen isotope composition of organically and conventionally grown crops. *Journal of Agricultural and Food Chemistry* 55: 2664–70.

7   Hohmann, M., Christoph, N., Wachter, H. & Holzgrabe, U. 2014. ^{1}H NMR profiling as an approach to differentiate conventionally and organically grown tomatoes. *Journal of Agricultural and Food Chemistry* 62: 8530–40.

Chapter 10

1   *On Smell* is a short essay by Lewis Thomas, published in the book *The Bedford Reader*. 1985. St Martin's Press, New York.

2   Laska, M., Seibt, A. & Weber, A. 2000. 'Microsmatic' primates revisited: olfactory sensitivity in the squirrel monkey. *Chemical Senses* 25: 47–53.

3   Shepherd, G. M. 2004. The human sense of smell: are we better than we think? *PLoS Biology* 2(5): e146.

4   Deibler, K. D. & Delwiche, J. (eds) 2004. *Handbook of Flavor Characterization: Sensory Analysis, Chemistry, and Physiology*. Marcel Dekker, Inc., Monticello, New York, p. 187.

5   Olsson, M. J., Lundström, J. N., Kimball, B. A.,
    Gordon, A.R., Karshikoff, B., Hosseini, N., Sorjonen,
    K., Höglund, C.O., Solares, C., Soop, A., Axelsson, J.
    & Lekander, M. 2014. The scent of disease: human
    body odor contains an early chemosensory cue of
    sickness. *Psychological Science*: 0956797613515681.

Acknowledgements

We would like to thank our publisher, Jim Martin, first for liking our idea and then for his patience as we took our time to develop it. We are grateful to our editor Catherine Best for all of her improvements to the manuscript and to Anna MacDiarmid for keeping us organised in the final stages of publication. We would also like to thank the many people we have spoken to over the last two years who have provided us with information, shared their viewpoints and shaped our own thinking about food fraud. Finally, to the many people out there – from government analysts to forensic accountants – who are fighting the good fight and trying to keep one step ahead of the fraudsters, we are grateful for all that you do.

Richard's acknowledgements

Chemists tend to write scientific papers rather than books but there are occasions when we need to express our thoughts and feelings on a subject in a different way. I will be forever indebted to Nicola for agreeing to work with me on this book – she has been absolutely amazing! The project has worked because we share a deep belief in *proper food* and a collective horror and indignation that anyone involved in supplying food to people would not want it to be *genuine* and *high quality*. She is generous in acknowledging me for the idea but this book would never have happened without her. I cannot thank her enough for her energy and creativity. She has been 'the boss' in this project and has to take all the credit for organising our publisher and polishing my final drafts. It was quite a jump for me from conventional scientific writing to a popular work – I've learnt a huge amount from her.

When I look back on my career I cannot thank my late parents enough for all the encouragement they gave me as a child to pursue my favourite interests. Parents often pass on critical skills, knowledge and views on life in less obvious ways. Boys in the 1960s weren't supposed be interested in cooking but I vividly remember sitting in the kitchen transfixed by my mother putting together the meals for the family from *raw ingredients*. She was unwittingly simultaneously inspiring my interests in food and chemistry – for cooking is chemistry! My father passed on to me his engineering skills – much of the analytical chemistry I do is about instrumentation and engineering. They were also impeccably honest people, which is undoubtedly why the whole idea of food fraud really rankles me.

All the pieces of the jigsaw puzzle were in place and so it's no wonder that one day my research would begin to challenge food fraud. I was lucky to work with Barry Rossell on the Leatherhead Food Research Association maize oil adulteration project; he gave me vitally important insights into how the food industry worked. Our PhD student Simon Woodbury did exceptional work and in the end we could tell you whether the corn oil in your kitchen was corn oil! I would like to thank Tim Knowles for editorial work on early drafts and Ian Bull for checking all the structures. I also have to thank my darling wife Gosia for encouraging me to work on the book and reading my drafts – she said some of it was 'quite interesting' so I'm optimistic others might think it's worth a read. Finally, I can only apologise to all my family and friends who have had to listen to me recounting endless food fraud tales and horror stories for the past couple of years.

Nicola's acknowledgements

I would first like to thank my co-author Richard for having the idea to write this book and then for helping to see the idea through. I am grateful that he is skilled in making complex topics accessible, otherwise this endeavour would have been far more challenging. I am indebted to the Knight Science Journalism Program at the Massachusetts

Institute of Technology for awarding me a fellowship to indulge in a week of learning about the science of food. Thank you to my editors – Ian Glen (aka Dad), Jennifer Gruno, Shelby Temple, Marlayne Glen, Leslie-Ann Glen and Amanda Woodman-Hardy – for their insightful comments on earlier versions of some of the chapters. I would like to acknowledge Dr Tom E. Reimchen, who not only taught me to think critically but has also, over the years, engaged me in many discussions on the ethics of what we eat. To my community that has supported me in various ways throughout this process: Louise Slade, Sandra Banner, Linda Oglov, Steve Simpson, Alida Robey, Paola Spivach, Claire Matthews, Sally Miller, Chris Slade, Helen Roberts, Nick Roberts, Katie Martin and Jennifer Kingsley – thank you. You have been essential to my productivity and my sanity and I apologise for sharing all of my research with you – I hope you will one day forgive me and find the will to eat again. My sister, Jennifer Gruno, has always been an inspiration to me in her commitment to eating mindfully – thank you. I am eternally grateful to my mum, Valerie Tregillus, for thinking anything I write is wonderful (which it's not) and for raising me on wholesome home-grown food. Though I may have complained about the dense brown bread sandwiches and straight-from-the-cow milk as a child, it set me up for a lifelong appreciation and understanding of good food. Last, but certainly not least, I am thankful to my supportive family. My husband, Shelby, and our son, Morgan, have been there when I needed them and have made themselves scarce when I didn't. Thank you.

Index

ABP Food Group 151
Accum, Frederick 51, 52, 171,
    204–205, 216, 270
  *A Treatise on Adulterations of
    Food, and Culinary
    Poisons* 49–50, 192
ACGT Inc. 136
adulteration (definition) 17–18,
    24, 49–50
  adulterated spices 206–15
  fruit products 257
  pepper 203–206
alcohol 250–53
allergies 127
  allergen regulations 221–22
  nut allergy 15, 210, 220–21
Allied Crude Vegetable Oil
    Refining Co. 85–86
anal leakage 127
Anchor Seafood Inc. 128
Angelis, Tino De 85–86
aniline 91–92
animal welfare 170–71
Argentina 147
Auramine-O 209
Australia 120, 147
  fruit exports 255–56
  wine 236, 238
Australian Marine Conservation
    Society 129
Austria 21, 236–37

Baltimore Sun 128
Barcode of Life Database
    (BOLD) 76, 78, 134
barcoding 75–78
  seafood 115–18
  taking DNA barcoding to the
    market 118–22

Bart Ingredients 221
BBC 142, 198
Beck, Lewis Caleb *Adulterations of
    Various Substances Used in
    Medicine and the Arts* 50
beef 114
  beefburgers 149–51
  fake beef 158
Bennett, Caroline 138
Bihar Times 190
blood products 108, 163–64, 169
Blythman, Joanna 262
Boston Globe 130
Bradley, Henry W. 181
bread 46
British Retail Consortium
    (BRC) 45
Broadbent, Michael 226–27
BSE (bovine spongiform
    encephalopathy)
    107–108, 168
Burger King 145
butter 181
  butter substitutes 182–84
  I can't believe it's not butter
    184–86
buying food 282–83
  buy from people you trust
    283–84
  buy locally produced
    food 283
  buy realistically priced
    food 284
  buy whole, recognisable
    foods 283
  expand your internal food
    reference database 284–85
  finding the story in your
    food 285–86

Cadbury's   169
Campden BRI   235–36, 239
Canada   108, 109, 112–13,
    129–30, 264
  seafood controls   133, 134–6
Canadian Food Inspection
    Agency (CFIA)   38–39, 271
carbon dating   227–28
cardamom   222–23
catfish   29–30, 113
cayenne   207
CGI (computer-generated
    imagery)   12
cheap food   33–35
cheese   176–77
  I can't believe it's not
    cheese   186–90
  processed cheese   262–63
chicken   114
  condemned meat   139–40, 141
  free-range chicken   166
  hydrolysed protein   159–61
chillies and chilli powder   208
China   10, 39
  China Food and Drug
    Administration (CFDA)   196
  citrus fruit   255–56
  criminal penalties for food
    fraud   41
  Fuyang formula fraud   195–96
  honey exports   83
  meat consumption   147
  Melamine Scandal   15, 17, 20, 41,
    52, 196–200, 270–71
  rice consumption   31–32
  sheep-free mutton   24, 154–55,
    157–58
chocolate   169
Christie's   226–27, 230
chromatography   217–18
  HPLC (high-performance liquid
    chromatography)   65–66, 72,
    74, 185, 196–97, 217, 246–47, 270

cinnamon   208
climate change   30–33, 286
  spices   222–23
Co-operative   55
Coates, Lewis   142
cod   32, 112–13, 120
Codex Alimentarius   63, 64, 97,
    99, 111, 194, 243
colourants   211–12, 213
Comigel   151
condemned meat   36, 139–43
  horse   149–53
  meat substitutions   157–61
  minced and processed meats
    153–57
  rotten history   143–47
  tight reins   147–49
consumers   45–47
  buying food   282–86
  fish identification   136–37
coriander   208–209
corn syrup   13–14, 66–68, 71, 244
cow dung   209, 217
Creutsfeld-Jakob disease
    (CJD)   107
criminal penalties for food
    fraud   40–42, 140–42,
    170, 231
cumin   15, 55–56, 209–10

dairy products   15
  I can't believe it's not
    cheese   186–90
  in search of dairy substitutes
    180–84
  our special relationship with
    milk   174–78
  what is milk made of?   178–80
  I can't believe it's not butter
    184–86
David Suzuki Foundation   129
Denby Poultry Products   139–40
Denmark   36–37, 42, 45

detecting fraud   49–51
   DNA: the ultimate fingerprinting
      approach?   75–78
   framing the food fraud
      question   57–59
   game changes   81–84
   guiding principles of food fraud
      detection   59–62
   honey case study   62–64
   how pure is our honey?   64–66
   how things have changed   51–54
   mānuka mystery   69–70
   metabolomics   73–75
   'omics' and the food testing
      revolution   70–71
   potential of proteomics   71–73
   stable isotopes to the rescue!
      67–68
   where does my honey come
      from?   78–81
   who's doing the testing?   54–57
diethylene glycol   21, 236–37
DNA testing   23, 47, 70–71
   barcoding life   115–18
   beefburgers   149–51
   DNA: the ultimate fingerprinting
      approach?   75–78
   spices   220
doner meat   157
donkey meat   155, 157, 177
Doyle, Scott   128

e-noses   103
e-tongues   103
eggs   9
   fake eggs   9–11
electronic sensors   103
ELISA (Enzyme-Linked
      ImmunoSorbent Assay)   154,
      156–57, 160, 221
Elliott, Chris   35, 36, 38, 41–42,
      152, 169
Elroy, Jim   228–29

environment   127–30
Epidemiological Review   91
essential fatty acids   89
European Food Safety Authority
      (EFSA)   155, 165
European Union (EU)   35–36,
      39, 262
   allergen regulations   221–22
   FishPopTrace   121
   food traceability project
      (TRACE)   79–81
   olive oil labelling   102–103,
      105–106
   seafood controls   128–29, 132,
      134–35
Expo Foods Ltd   40

fatty acids   88–90
Findus   150
fingerprinting   59–62
   chemical complexity of foods   60
   DNA: the ultimate fingerprinting
      approach?   75–78
   food processing   61
   how pure is our honey?   64–66
   inherent variability   60
   metabolomics   73–75
   'omics' and the food testing
      revolution   70–71
   potential of proteomics   71–73
   stable isotopes to the rescue!
      67–68
fish   32–33, 108
   barcoding life   115–18
   dealing with diversity   114–15
   farmed fish passed off
      as wild   122–25
   fish laundering is bad for the
      environment   127–30
   fishing industry   109–10
   short-weighting   130–32
   susceptibility to
      pseudonyms   110–14

'swordfish' with a side of anal
    leakage   125–27
  traceability   135–36
Fish Barcode of Life
    (FISH-BOL)   118–22
FishPopTrace   121
Food and Agriculture
    Organization (FAO)   63
Food and Safety Standards
    Authority of India
    (FSSAI)   18, 191, 193
food fraud   6–7, 12–16, 286–88
  can we blame government?
    37–42
  cheap food conundrum   33–35
  defining the problem   17–22
  fraud forecast   22–23
  horse meat scandal   35–37
  responsibility of industry   42–45
  role of science   47–48
  tipping the scales on fraud   25–30
  traditions of trickery   23–24
  uncertain futures   30–33
  watchdog organisations and
    watchful consumers   45–47
food inspection   26–27
food miles   27–28, 111, 113–14
foodborne illnesses   21, 27, 146
Forbes   227
Forbes, Christopher   227
formalin   16, 51, 191–92,
    258–60, 271
Frank Leslie's Illustrated
    Newspaper   173
Frankel, Edwin   101–103
Frericks, Hans-Peter   227–28
fruit   16, 52, 255–56
  fruit juice   241–50
  on the origin of veggies
    262–66
  preservation   256–62
  production perjuries   266–69
  rotten apples   271–72

Gandhi, Maneka   191
Gardner, A. K.   173
GenBank   78, 116, 118
Germany   236–37
Gilbert, Tom   77
ginger   210
Goodman, Larry   151
grains   16
grass-fed livestock   165–66
Greenberg, Eric   229
grouper   29–30, 113
Guardian   27, 45, 106, 151, 188

haddock   32, 33, 120
halal meat   159
halibut   116, 119, 121, 129–30
ham   161–62
Hanner, Robert   118–20, 122,
    125, 281
Hassall, Arthur Hill   50, 204–205,
    206, 208, 210
Hebert, Paul   75, 117–18
HFCS (high fructose corn
    syrup)   66–68, 71, 244
Hitchin' Post Steak Co.   40–41
honey   13–14, 83–84
  barcoding honey   78
  honey case study   62–64
  how pure is our honey?   64–66
  mānuka mystery   69–70
  metabolomics   73–75
  proteomics   72–73
  stable isotopes to the rescue!
    67–68
  where does my honey come
    from?   78–81
Hong Kong Centre for Food
    Safety (CFS)   199–200
horse meat   11, 15, 39–40, 153
Horsegate   23, 28, 40, 53, 55, 56,
    149–53, 168, 286
  effect of Horsegate   169–72
  effect on governments   35–37

HPLC (high-performance liquid
       chromatography)   65–66,
       72, 74, 185, 196–97, 217,
       246–47, 270
Hubert, Philippe   228
hydrolysed protein   159–61
    milk   199–200

Independent   226
India   15, 21, 133, 218
    I can't believe it's not milk
       190–93
International Olive Council
       (IOC)   98, 101–102
isotopes   67–68, 80–81, 265–66
    organic produce   267–69
Italy   21
    olive oil production   104–106
    Tropea red onions   262–63
    wine   237, 239

Jack, Lisa   47
Jefferson bottles   226–31
juices   19, 241–50
    juice jargon   242–43
    plasticisers   21, 248–9
    pulp wash   246–47

Kennedy, John F.   85
KFC   145
Kircher, Amy   22–23, 27
Koch, Bill   227–31
KPMG   42
Kurniawan, Rudy   230–31
Kwik Save   140

labelling   53–54
    olive oil labelling   102–103,
       105–106
    production perjuries   266–69
Labra, Massimo   78
lactose   176, 179, 186, 192
Lancet   45, 50

Lawrence, Felicity   27–28, 45
    Not on the Label   262
lead poisoning   24, 234
lettuce   30–31, 33, 267,
       269, 285
Linden Foods   43
local produce   29–30
Luxembourg   147

mad cow disease   107–108
maize oil   27, 94–97
malachite green   124, 208, 209
Malthus, Thomas Robert   171
mangoes   16, 258–59
margarine   182–84
Marine Conservation Society
       (MCS)   20, 129–30, 137
Marko, Peter   116–17, 119–20
Marks & Spencer   112
Martina, Maurizio   105
McDonald's   43, 88, 145
meat   19–20, 114
    effect of Horsegate   169–72
    meat consumption   147
    meat glue   164–65
    mischievous meat manipulations
       165–67
    motivations and implications of
       meat fraud   167–69
    religious restrictions   159,
       168–69
    shaped meats   161–65
    substitutions   153–57
    substitutions aren't limited to
       processed meat   153–57
    see condemned meat
Mège-Mouriès, Hippolyte   181
metabolomics   73–75, 266–67
metanil   21, 213, 216, 218, 270
microscopy   217
milk   174–78
    formula fraud   193–200
    hydrolysed protein   199–200

I can't believe it's not butter
    184–86
I can't believe it's not
    cheese 186–90
I can't believe it's not milk
    190–93
in search of dairy substitutes
    180–84
melamine milk 17, 19, 20, 41,
    52, 196–200, 270–71
swill milk 173–74
urea 21, 191–93, 225
what is milk made of?
    178–80
minced meat 155–56
MK Poultry 140
monkfish 125–27, 136–37
Monterey Bay Aquarium Seafood
    Watch 20, 129
Monty Python 186, 285
Moon Fishery India Pvt.
    Limited 133
Moshi Moshi 137–38
Mueller, Tom Extra Virginity
    58, 106
Mullaly, John 174
mutton 114
    fake mutton 24, 154–55,
    157–58

nanoparticles 260–61, 287
Napoleon III of France 181
Nature 116
Neal, Jillian 210
Neill, Charles P. 144
Neptune Fisheries Inc. 128
Netherlands Food and Consumer
    Product Safety Authority
    (NVWA) 39
New York Times 24, 122, 173,
    174, 195–96
New Yorker 226
New Zealand 120, 133, 147

New Zealand Unique Mānuka
    Factor Honey Association
    (UMFHA) 69
Norwest Foods International
    Ltd 151–52
NSF International 44
nut allergy 15, 210, 220–21
nutmeg 202–203, 210–11

Obama, Barack 132
Oceana 20, 45, 119–20, 137
oilfish 127
olive oil 93–94
    grades of olive oil 98–99
    likely victim of food fraud
    97–101
    olive oil hits the fan 101–104
    perfect storm 104–106
onions 262–63
Opie, Andrew 45
organic produce 53, 266–69
OSI Food Solutions 43

Papa John's 145
paprika 15
parasites 32, 123, 146, 149
    farmed fish 124–25
Pasteur, Louis 235
Pauly, Daniel 110
pepper 202–203
    pepper is not all it's cracked
    up to be 203–206
pesticide residues 264–65
Phylloxera 25–26, 234
Pizza Hut 145
plasma 108, 164
plasticisers 21, 248–9
Poivre, Pierre 202
POM Wonderful 19
population growth 171
pork 114
    fake beef 158
    in halal meat 159

poultry 40–41, 114–15
  poultry by-products 141–43
preserving 256–62
processed foods 52–53, 61
processed meat 154–55
protected geographical indication
  (PGI) 262–64
proteomics 71–73, 153–54
puffer fish 125–26
pulses 16, 270–71
Purely Juice Inc. 19

raisin wine 25–26, 234–35
rapeseed oil 91–92
Red Lion abattoir 151
red mullet 32, 119
red snapper 20, 110–11, 116–17,
  119–20
Reily Foods Company 55–56
Renton, Alex 106
responsibility of food
  industry 42–45
Reynolds, James Bronson 144
rice 31–32, 269–70
Righetti, Pier Giorgio 72–73
Riverford Organics 285
Roberts, Peter (Maggot
  Pete) 139–41
Rodenstock, Hardy 226–31
Roosevelt, Theodore 144
Rossell, Barry 95–96
Royal Botanic Gardens, Kew 78
Royal Society 13

saffron 211–12, 216–17, 217–18
Sainsbury's 55, 140
salmon 33, 111, 121–24, 137
Salmonella 21, 133, 146
salt 212
Sanlu Group Co. Ltd 19, 41
sausages 15, 40, 154, 156–58, 164,
  168–69
scallops 130–31

science 47–48
Scotch Whisky 252–53
seafood 14–15, 20, 108
  barcoding life 115–18
  catfish–grouper scandal 29–30
  climate change 32–33
  closing the net on seafood fraud
    132–38
  dealing with diversity 114–15
  farmed fish passed off as
    wild 122–25
  fish laundering is bad for the
    environment 127–30
  fish out of water 130–32
  susceptibility to pseudonyms
    109–14
  'swordfish' with a side of anal
    leakage 125–27
  taking DNA barcoding to the
    market 118–22
Selten, Willy 151, 170
senses 273
  smell 46–47, 273–76
  taste 277–79
  technology to the rescue? 279–82
Shanghai Husi Food Co. 145
Shanken, Marvin 227
sharks 120, 128
Silvercrest 151–52
Sinclair, Upton The Jungle 144
smell 46–47, 273–76
sodium tripolyphosphate
  (STPP) 130–32
Soybean Scandal 85–86, 90, 287
Spain 14, 20–21, 29
  Toxic Oil Syndrome
    (TOS) 90–93, 97–98
spectroscopy 218–19
spices 15, 201–203
  adulterated spices 206–15
  finding the fakes 215–24
  pepper is not all it's cracked up
    to be 203–206

spirits   250–53
Starbucks   145
strawberry flavour   46
Sudan dyes   208, 213, 218, 255
supply chains   28–29, 42–
    auditing   42–45
    seafood   111–12
swill milk   173–74
Switzerland Cheese Marketing
    Association   190
swordfish   127, 128, 137

Taiwan   41, 248–50
taste   277–79
Taylor, Geoff   235–35, 239–40
technology to the
    rescue?   279–82
Tesco   28, 44, 55, 140, 149–50, 152
testing foods   54–57
    country-of-origin testing   166–67
    finding the fake spices   215–24
    'omics' and the food testing
        revolution   70–71
    on the origin of veggies   262–66
    see DNA testing
ThisFish Database   135, 137
Thomas, Lewis   273
tilapia   20, 119
Tirth, Swami Achyutanand   193
tomatoes   20–21, 53, 59, 92–93
tonka beans   213–24
Tradex Foods Inc.   136
transglutaminase   108, 163–64
tuna   33, 119, 128, 133
turkey   114–15
turmeric   212–13, 21

UK   19–20, 21, 43–44
    cheap food   34–35
    food imports   28–29
    Food Integrity Committee   38
    seafood consumption   32–33
    vCJD   107

UK Department for Environment,
    Food and Rural Affairs
    (Defra)   27, 38
UK Food Crime Unit   35, 37
UK Food Standards Agency (FSA)
    17–18, 37–38, 44, 56, 69, 97
    fake vodka   250–51
    food hygiene   146–47, 148–49
    hydrolysed protein   159–61
UK Ministry of Agriculture,
    Fisheries and Food
    (MAFF)   27, 95
US   14, 17, 21, 147
    cheap food   33–34
    criminal penalties for food
        fraud   40–41
    honey imports   83
    National Center for Food
        Protection and Defense
        (NCFPD)   22–23, 33, 286
    Salad Oil Scandal   85–86, 90, 287
    seafood controls   132–3
US Food and Drug
    Administration (FDA)
    17, 26–27, 35, 38, 56
    fish species   110, 113, 114,
        120–22, 125–26, 133–34
    food hygiene   146, 148
US International Trade
    Committee (ITC)   29–30
US National Marine Fisheries
    Service (NMFS)   112
US National Oceanic and
    Atmospheric Administration
    (NOAA)   128, 134
US National Seafood Inspection
    Laboratory (NSIL)   110
US Pharmacopeia (USP) Food Fraud
    Database   39, 104, 153, 207

vanilla   213–15
vCJD (variant Creutsfeld-Jakob
    disease)   107, 168

vegetable oil   14, 20–21,
    85–86
  big money, big cheats   86
  catastrophic vegetable oil
    fraud   90–93
  cracking vegetable oil
    fraud   94–97
  getting clear on oil   86–90
  hydrogenation   181–84
  oil identity crisis   93–94
  olive oil   97–101
  olive oil hits the fan   101–104
  perfect storm   104–106
vegetables   16, 52, 255–56
  on the origin of veggies
    262–66
  preservation   256–62
  production perjuries   266–69
  pulses, grains and seedy
    transactions   269–71
Vitter, David   134
vodka, fake   250–52
Von Liebig, Justin   194
Vonderplanitz, Aajonus   146

Waitrose   44
Washington, George   226
watchdog organisations   45–47
Wen Jiabao   195
Which?   45, 56
whisky, fake   251–53
Wiley, Harvey Washington   51
Willy Selten BV   151–52
Wilson, Bee *Swindled*   17
wine   15–16, 21
  holistic approach to wine crime
    237–41
  Phylloxera   25–26, 234
  wine fraud   225–31
  wine making   232–37
Wine Spectator   227
Wong, Eugene   118–19
Woodbury, Simon   96–97
World Health Organization
    (WHO)   90–91
World Wide Fund for Nature
    (WWF)   129

Zachys auction house   229–30